The Outdoor Lighting Guide

T0174278

The Institution of Lighting Engineers (ILE) is the UK and Ireland's largest and most influential professional lighting association, dedicated solely to excellence in lighting. Founded in 1924 as the Association of Public Lighting Engineers and licensed by the Engineering Council, the ILE has evolved to include lighting designers, architects, consultants and engineers amongst its 2,500-strong membership.

The Outdoor Lighting Guide

The Institution of Lighting Engineers

Routledge
Taylor & Francis Group

LONDON AND NEW YORK

First published 2005 by Routledge

2 Park Square, Milton Park, Abingdon, Oxfordshire OX14 4RN
52 Vanderbilt Avenue, New York, NY 10017

Routledge is an imprint of the Taylor & Francis Group, an informa business

First issued in paperback 2019

Typeset in Sabon by
Integra Software Services Pvt. Ltd, Pondicherry, India

British Library Cataloguing in Publication Data
A catalogue record for this book is available
from the British Library

Library of Congress Cataloging in Publication Data
A catalog record for this book has been requested

ISBN: 978-0-415-37007-3 (hbk)
ISBN: 978-0-367-39178-2 (pbk)

Contents

Foreword

Outdoor lighting is used for a variety of purposes in modern society. It enables people to see essential detail so they can be active at night. Good lighting can enhance the safety and security of persons or property, emphasise features of architectural or historical significance, or call attention to commercial premises by means of area lighting or signs. Unfortunately, poor lighting practice is extensive. Much bad lighting can be blamed on the fact that the user is unaware of the issues of visibility and its usefulness. Careless and excessive use of artificial light in our outdoor environments causes extensive damage to the aesthetics of the night-time environment, while at the same time it often compromises safety and usefulness, the very reason for its installation. Bad lighting hurts everyone. The loss of the dark starfilled sky is of tragic consequence for the environment and for the human soul, akin to the loss of our forested landscapes and other natural treasures. On the other hand, quality lighting brings substantial benefits. Lack of glare and excessive contrast brings improved visibility, especially for the ageing eye. Elimination of wasted light saves money, energy and resources, which in turn reduces air pollution and carbon dioxide emissions caused by energy production and resource extraction. Quality lighting improves the appearance of our communities, returning a sense of balance to the night and giving a more attractive appearance to our cities, towns and villages.

So good lighting can make a significant contribution to the outdoor environment whereas poor lighting can damage it.

This positive contribution is not limited to the hours of darkness, as the reduction in crime effects are now known to extend to the daytime. In August 2002 the British Home Office published two research studies on crime prevention: 'Effects of improved street lighting on crime: A systematic review', and HORS 251 'Crime prevention effects of closed circuit television: A systematic review', HORS 252.[1]

In international experiments one of the main points to emerge from the street lighting study is that where street lighting had been improved there had been an overall reduction in recorded crime of 20 per cent. In the British studies there was a 30 per cent decrease in crime. The authors conclude that lighting improvements increase community pride and confidence and strengthen informal social control, and that this explains the impact, rather than increased surveillance or deterrent effects. Furthermore, improvements in street lighting offer a cost-effective crime reduction measure.

The Closed-Circuit Television (CCTV) study summarised the findings of previous studies from both Britain and the USA and concluded that where CCTV had been installed there had been an overall reduction in recorded crime of 4 per cent across all the experimental areas. It was found that CCTV had no effect on violent crimes but had a significant desirable effect on vehicle crimes.

Both studies together demonstrate that improved lighting is between five and seven times more effective at reducing crime than the installation of CCTV.

In 1999 the Technical Committee of the Institution, then under the chairmanship of Stuart Bulmer, recognised the need to bring together various elements of good outdoor lighting practice. Some were already contained within Institution documents but others were not yet committed to paper. A panel of experts was set up under the leadership of the incoming Technical Committee chairman David McNair to cover the range of topics. The aim of the project was to produce a comprehensive outdoor lighting guide. It was to be a one-stop shop: able to suitably educate any engineer (of whatever persuasion) or other related professional who cares to examine it; and, to document current good practice, target lighting levels, uniformity and glare control required for the different applications. It was to be comprehensive yet only explain what is necessary, with reference made to further documents to assist those with more expertise.

The panel contributed in their field of expertise to the various Parts of the document. As expected there were multiple overlaps between the sections and the authors.

The panel members were as follows:

John S. Anderson	Whitecroft Road and Tunnel Lighting
John Brewis	Institution of Lighting Engineers
David S. Black	South Lanarkshire Council
David Burton	Ashfield Consultancy Services
David Coatham	Institution of Lighting Engineers
Jason Ditton	University of Sheffield
Robert Divall	Thorn Lighting
Allan Howard	Mouchel Consulting
Carl Gardner	Institution of Lighting Engineers
Arthur Gibbons	Dron and Dickson Group
Ian Graves	Philips Lighting
Clive Lane	CU Phosco Lighting
William Marques	CU Phosco Lighting
David G. McNair (Chairman)	South Lanarkshire Council
Nigel Parry	City of Westminster

Nigel E. Pollard NEP Lighting Consultancy
Gareth Pritchard Pudsey Diamond Engineering Ltd
Malcolm Richards D.W. Windsor
Bryan Shortreed Urbis Lighting Ltd

The Institution is grateful for the support of the contributors and their affiliated bodies, without whom the compilation of this book would not have been possible.

The Chairman, working to the guidance of a strategy group, drew together the various contributions into what hopefully readers will find a coherent structure. This would not have been possible without the extensive and invaluable effort of David Coatham.

The strategy group members were as follows:

Patrick Baldrey Urbis Lighting Ltd
David Coatham Institution of Lighting Engineers
David McNair South Lanarkshire Council
Gareth Pritchard Pudsey Diamond Engineering Ltd
Derek Rogers Derek Rodgers and Associates

For final checking John Brewis joined the strategy group. In addition the Institution wishes to acknowledge the assistance received from Steve Lain and Roger Heyworth, who started but were unable to complete sections due to changes in their employment; Alastair Scott of Urbis Lighting, who provided extensive and invaluable assistance in accessing diagrams and photographs; Richard Leonard of Philips Lighting, who carried out many calculations; and Colin Rowley of the Dron and Dickson Group, who was too busy to contribute, but commissioned Arthur Gibbons to work in his place. CU Lighting and South Lanarkshire Council provided diagrams. Philips Lighting, South Lanarkshire Council, Thorn Lighting, D.W. Windsor and Urbis Lighting provided photographs. Extracts from the work of Ken Pease and the International Dark-Sky Association are gratefully acknowledged, as are the words added by Kate Painter. The 'Slipstream' sculpture in Chapter 4 is by Joseph Ingleby and the photographs of it by Ruth Clark. The book was prepared for publication by David Coatham.

It is hoped this book will give readers a better understanding of the principles of good lighting, act as a reference source where good practice can be identified and in some cases explain the limitations under which designers work.

Finally, this document is a building block for the future. As techniques develop, as knowledge is acquired and as experience is gained it will have to be updated. If readers have any comments, suggestions for improvement, requests for inclusion of contributions towards future editions or wish to view what others are posting, they are invited to visit www.ile.org.uk.

David McNair

Disclaimer

This publication has been prepared on behalf of the ILE Technical Committee for study and application. The document reports on current knowledge and experience within the specific fields of light and lighting described and is intended to be used by the ILE membership and other interested parties. It should be noted, however, that the status of this document is advisory and not mandatory. The views expressed are not necessarily those of the contributors. The ILE should be consulted regarding possible subsequent amendments. Any mention of organisations or products does not imply endorsement by the ILE. Whilst every care has been taken in the compilation of any lists, up to the time of going to press, these may not be comprehensive. Compliance with any recommendations does not itself confer immunity from legal obligations.

The Institution of Lighting Engineers

The objective of the Institution is to promote, encourage and improve the science and art of lighting for the benefit of the public.

Abbreviations and symbols for quantities, units and notation

A	Ampere
ABS	Acrylonitrile-butadiene-styrene
BS EN	British Standard European Norm
BSI	British Standards Institute
cd	Candela
CE	European conformity mark
CIE	Commission Internationale de l'Eclairage
CRI	Colour rendering index
CCT	Correlated colour temperature
DLOR	Downward light output ratio
E	Illuminance
$\bar{E}$	Average illuminance
EEC	European Economic Community
EMC	Electromagnetic compatibility
EN	European Norm
ENEC	European Norm electrical certificate
EPDM	Ethylene propylene diene monomer
GLS	General lighting service
GRP	Glass-reinforced polyester
HID	High intensity discharge
HPS	High pressure sodium
HQI	Metal Halide
Hz	Hertz
IP	Ingress protection
IR	Infrared
I	Luminous intensity
I table	Luminous intensity distribution table
ILCOS	International lamp coding system
K	Kelvin
L	Luminance
$\bar{L}$	Average luminance
LDL	Lighting design lumens

lm	Lumen
LOR	Light output ratio
lx	Lux
MICC	Mineral insulated copper conductors
MF	Maintenance factor
PAR	Parabolic reflector
PC	Polycarbonate
PCA	Polycrystalline alumina
PFC	Power factor correction capacitor
PIR	Passive infrared
PMMA	Acrylic (polymethylmethacrylate)
PVC	Polyvinylchloride
R table	Reflectance table
R_a	Colour rendering index
RAL	Reichs-Ausschus fur Lieferbedingungen
RCD	Residual current device
SIP	Superimposed pulse
SON	High pressure sodium
SOX	Low pressure sodium
sr	Steradian
TI	Threshold increment
ULOR	Upward light output ratio
ULR	Upward light ratio
UV	Ultraviolet
U_l	Longitudinal uniformity
U_o	Overall uniformity
V	Volt
W	Watt

Chapter 1

Visual effects of lighting

1.1 Introduction

Lighting has three primary functions:

1   To improve visual performance;
2   To improve safety; and
3   To improve the visual environment.

Each of these functions is equally applicable to both indoor and outdoor environment, however; the outdoor environment is a very different place from the indoor environment. It is subject to more extreme environmental conditions and human behaviour. Whilst there is some overlap of task there is a complete overlap of lighting principles and visual effects.

1.2 Light

1.2.1 Lighting levels

The human eye perceives objects by the light that is emitted or reflected by them. With the exception of light sources, the light reaching an observer (the luminance of the object) is the reflected light and is dependent on the light incident on the object (the illuminance), the reflective properties of the object and the position of the observer with relation to the object. These variables give a very large number of possible luminance requirements.

1.2.2 Task performance

The ability to perform a visual task is influenced by the size of the task, the contrast and the vision of the viewer. Whereas a difficult task cannot be made into an easy task, increasing the illuminance generally improves visual performance for a specified task. However, saturation occurs, and beyond a certain value any further increase is superfluous and results in an unnecessary use of energy. The point at which saturation occurs will be

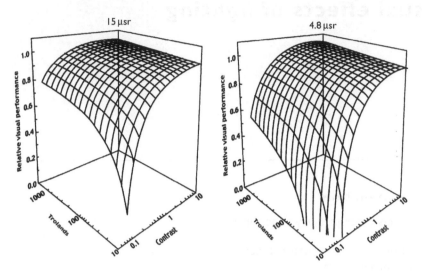

Figure 1.1 Relationship of relative visual performance with retinal illuminance in trolands and contrast, for task sizes of 15 μsr (a) and 4.8 μsr (b).

higher the more difficult the task e.g. tasks involving very small objects or those carried out at high speed. Saturation is illustrated in Figure 1.1.

1.2.3 Appearance

In a limited number of outdoor locations the appearance of the lighting will be more important than the task performed. Indeed the actual task may be viewing the appearance of a lit object or the lit environment. Examples are the illumination of buildings and structures and the lighting of town squares. The lighting here is intended to create mood, interpret architecture or give visual stimulation. The art of good lighting becomes as important as the science, and designers have to take account of colour, form, texture and perception.

1.3   Flux, intensity, illuminance, luminance and brightness

Flux is the total quantity of light that a source (e.g. a lamp) emits. It is measured in lumens and is the starting point for general lighting calculations.

The quantity of light emitted in a specified direction is the *luminous intensity* or simply the intensity of the light in that direction. The existence of luminous intensity diagrams or tables for luminaires allows detailed lighting calculations to be carried out. It is measured in candelas, which are lumens per unit solid angle in the specified direction.

Illuminance is the magnitude of light incident on a surface. It cannot be seen because it has not reached the eye yet. Illuminance is the objective

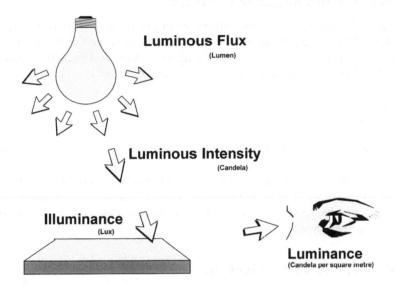

Figure 1.2 Flux, intensity, illuminance and luminance.

quantity that is most commonly used to specify lighting levels because in most applications it is not possible to specify the position of the observers, the lit objects and the light sources with sufficient accuracy to use luminance, or there are multiple observer positions. Illuminance is measured in lux, which are lumens incident on a point per area of the point.

However, lighting practitioners think in terms of luminance, contrast and glare. The amount of light that reaches the eye by reflection or by direct emission from a light source is called *luminance*. The light reflected from any surface is dependent on the quantity of illuminance, the reflective properties of the surface and the position of the observer with relation to the surface. Luminance is measured in candelas per square metre, which is the luminous intensity per area of the solid angle of the object when viewed by the observer.

Figure 1.2 illustrates all the four terms. *Brightness* is the subjective response created by the brain's interpretation of what the eye sees.

1.4 Glare

Glare occurs when one part of the visual scene is much brighter than the remainder. The most common causes of glare are inappropriate orientation of luminaires, and the poor selection of luminaire and mounting height combination. In a road environment, dipped vehicle headlamps can cause substantial glare even if the road is well lit. Glare impairs vision, causes discomfort and reduces task performance.

1.5   Positive and negative contrast

Contrast is the assessment of the difference in appearance of two or more parts of a field seen simultaneously or successively. It is the key to vision: if there is no contrast between an object and its background then the object will not be detected. The luminance contrast of an object is

$$C = \frac{L_2 - L_1}{L_1}$$

where L_2 is the luminance of the object; and L_1 is the luminance of the background.

Where the object is brighter than the background, there is a positive contrast and the object is seen by direct vision. And where the object is darker than the background, there is a negative contrast and the object is seen by silhouette vision.

1.6   Absorption and reflection

Any light falling on to a surface that is not reflected is either absorbed or transmitted through the object.

If the material does not transmit light, all non-reflected light disappears into the surface and is converted into heat. This is called *absorption*. The amount of absorption varies according to the angle of incidence, the colour of the light and the physical characteristics (colour, texture, density) of the material. Generally, for higher angles of incidence more light will be reflected.

The colour of a surface is dependent on the light reflected from it, for example a blue surface will reflect incident light in the blue wavelengths of the spectrum and absorb light with other wavelengths.

1.7   Radiation

Light forms part of a complex of physical phenomena included under the heading 'electromagnetic radiation'. It is therefore closely related to, for example, radio and TV signals, infrared (IR) and ultraviolet (UV) radiations, X-rays and other radiations. These emissions occur at different wavelengths. The major difference between light and these other phenomena is that humans and animals use a collection of the wavelengths to 'see' and this is called the visible spectrum. Some animals also use wavelengths in the IR or UV ranges to extend their range of vision.

'White' light is a collection of different wavelengths between approximately 380 and 780 nm, which in combination are perceived as white. Most lamps that emit a 'white' light do not emit a continuous spectrum of wavelengths but a series of wavelengths of different amplitude. Figure 1.3 shows

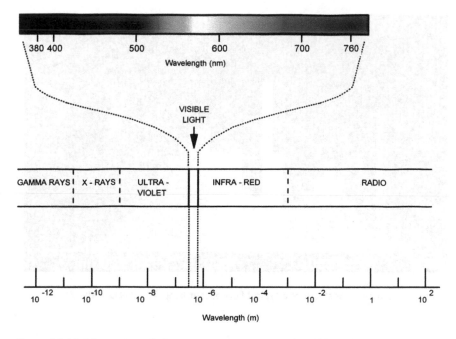

Figure 1.3 Visible portion of electromagnetic spectrum. *(see Colour Plate 1)*

the relationship between the visible portion of electromagnetic spectrum i.e. visible light and the non visible portions of the spectrum. Figure 1.4 shows the range and proportions of the emission spectrum of a typical fluorescent lamp.

1.8 Apparent colour

Colour temperature describes how a lamp appears when lit. It is the temperature of a black body radiator that emits radiation of the same chromaticity as the lamp being considered. For complete accuracy, the chromaticity must be on the black body (full radiator) locus, the power radiation curve of a black body.

As very few lamps have chromaticity on the locus, the more useful correlated colour temperature (CCT) is used. It is based on similar chromaticity to a black body radiator. This is red at 800 K, warm yellowish 'white' at 2800 K, daylight 'white' at 5000 K and bluish daylight 'white' at 8000 K. CCT is good for comparing incandescent lamps because they emit a continuous spectrum. It is not so good for comparing discharge lamps because their spectrum is not necessarily continuous. Some discharge lamps with the same CCT can have different effects on illuminated objects (Figure 1.5).

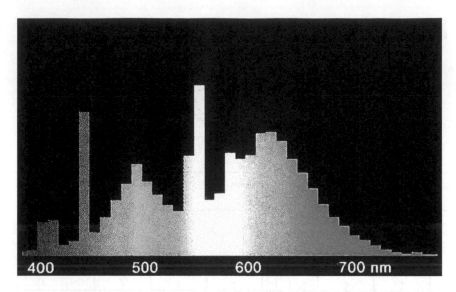

Figure 1.4 Emission spectrum of a typical fluorescent lamp. (see Colour Plate 2)

(a)

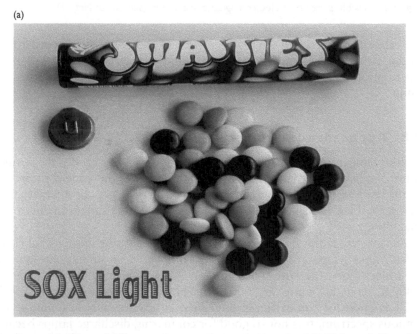

Figure 1.5 Effect of different colour temperature: (a) under SOX light and (b) under 'white' light. (see Colour Plate 3)

(b)

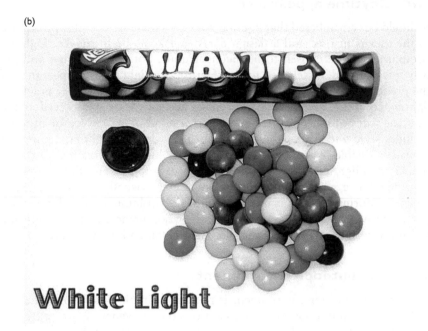

Figure 1.5 (Continued).

1.9 Colour rendering

The ability of a light source to render colours of surfaces correctly is quantified by the CIE Group and the colour rendering index (Table 1.1). This index is based on how close a set of test colours are reproduced by the lamp under evaluation, relative to how they are reproduced by an appropriate standard light source of the same CCT. Perfect matching is given value of 100.

Table 1.1 CIE colour rendering groups

Colour rendering groups	CIE general colour rendering index	Typical applications
IA	$R_a \geq 90$	Critical colour matching
IB	$90 \geq R_a \geq 80$	Accurate colour judgements required for appearance
2	$80 \leq R_a \leq 60$	Moderate colour rendering required
3	$60 \leq R_a \leq 40$	True colour recognition of little significance
4	$40 \leq R_a \leq 20$	Not recommended for colour matching

1.10   Daytime appearance

It should be remembered that lighting is a very visible service. Its appearance to the user is just as vital a design parameter as its performance. Lighting equipment can be chosen to be inconspicuous, both in its design and the way it is installed, or selected to be a feature. At night the lit appearance of the products used, including their supporting structures, whether brackets, lighting columns, towers or background, needs considering as well as the lit scene's appearance and technical performance. The choice of product includes style and scale, colour and finish. By day the lighting equipment is usually visible in the environment and the luminaire together with the bracket and lighting column or other means of support and fixing must be considered as a whole – they form a single item visually. What is seen, at night and during the day, is usually a collection of luminaires and their supports; the way these are seen together, their dominance, clutter, confusing or untidy combinations, for example, must be taken into account.

1.11   The outdoor environment

Lighting the outdoor environment is relatively simple during the hours of daylight as even in poor weather conditions the sky generally has sufficient luminance to provide adequate light for human observers who do not have significant visual impairment. Luminance is above $10\,cd\,m^{-2}$ allowing full operation of the cone receptors. Vision is in photopic mode with full colour differentiation. Relative sensitivity to different wavelengths is shown on the right-hand curve of Figure 1.6.

However, during the hours of darkness and where no artificial light is available, luminances drop to below $10^{-2}\,cd\,m^{-2}$. Vision is in the scotopic

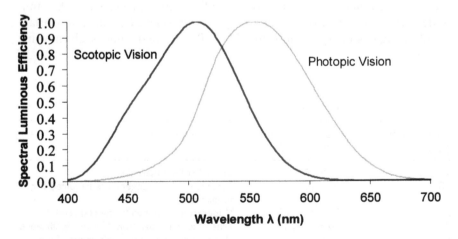

Figure 1.6 Spectral response of human eye.

range where only the rod receptors are operational, with no colour differentiation provided. Peripheral detection is superior to foveal detection, as the rods are concentrated in the peripheral areas of the eye. The peak sensitivity of the eye shifts to a lower wavelength of light. As a consequence the eye is relatively more sensitive to light in the blue areas of the spectrum. However, there is no colour sensation; all is grey. Relative sensitivity to different wavelengths is shown on the left-hand curve of Figure 1.6.

When artificial light is introduced to the field of view during the hours of darkness, luminances are generally between 10^{-2} cd m^{-2} and 10 cd m^{-2}. This mode of vision is called mesopic. As luminances increase, three effects occur: first, foveal detection becomes progressively more important than peripheral; second, colour becomes detectable; and third, the relative luminosity of the red wavelengths increases more strongly than that of the blue. Relative sensitivity of the eye to different wavelengths is represented by a curve that exists somewhere between the left- and right-hand curves of Figure 1.6. The relative sensitivity of the eye changes as the observer views different scenes in the mesopic range of vision.

As the light output of a lamp is determined by multiplying the power output at each wavelength by the relative sensitivity of the eye at that wavelength, this has implications for the choice of light source to be used in mesopic conditions.

Chapter 2

Social and environmental elements

2.1 Introduction

This section provides information on the context of lighting in society and the environment.

In addition to facilitating the safer movement of pedestrians and vehicles during the hours of darkness and enhancing commerce and recreation facilities, lighting has a key role in preventing crime and reducing the fear of crime. Crime prevention practitioners have always advocated improvements to outdoor lighting as part of their strategies; for example in Scotland powers to install street lights, including on private buildings, were enshrined in the Burgh Police (Scotland) Act 1896, which remained statute for almost a century. More recently, and after a lengthy debate, the UK Government in 2002 finally acknowledged the key role of lighting in reducing crime, with the publication of Home Office Research Paper 251, 'Effects of improved street lighting on crime: A systematic review'. An analysis is presented in Section 2.2.

Over the past 100 years advances in technology and decreases in costs have greatly increased the ability of lighting practitioners to deliver more light and society to afford it. However, a downside has been a proliferation of inappropriate lighting installations, which create sky glow and light pollution, and waste precious resources by the inefficient use of energy. Problems and solutions are described in Section 2.3.

Recognition of these issues has resulted in the formulation of light plans, an essential tool to achieve visual unification of the disparate night-time components of urban centres, to resolve conflicts between the lighting needs of different users, to aid the use of light as a commercial tool and to minimise light pollution. Essential information on light plans is included in Section 2.4.

There is a legal requirement and moral responsibility to ensure health and safety, and as it is impossible to cover all legislation, a comprehensive selection of relevant information is included in Section 2.5.

As the twin realisations that the earth's resources are limited and that inappropriate disposal of materials can be harmful to living organisms develop, so the importance of waste management increases. The principal lighting-related issues are examined and a review of waste management legislation is given in Section 2.6.

2.2 Crime and disorder

2.2.1 Introduction

Crime prevention practitioners have always included improvements to outdoor (particularly street) lighting in their toolbox and have repeatedly advocated its use.[1] However, over the last 15 years, the view that improved street lighting does not reduce crime has emerged. This view has been attributed to the Home Office, with the result that many have taken it to be an official position. If left unchallenged, this view would have the effect of excluding or limiting the role of improved street lighting in the local crime and disorder prevention strategies that are required under the Crime and Disorder Act 1998. It was thus thought timely to consider afresh the effect of street lighting on crime. The Head of the Home Office Crime Prevention Agency, Chief Constable Richard Childs, thus asked Professor Ken Pease to review the most up-to-date research evidence and this was done in July–August 1998. That review effectively overturned the conventional assumption that improvements to street lighting do not make a considerable impact on both crime and the fear of crime. What follows here is an attempt to summarise the general thrust of that review[2] for those involved in the lighting industry.

2.2.2 The effects of street lighting on crime and disorder

The basic conclusions of the review are as follows.

Precisely targeted improvements to street lighting generally have crime reduction consequences, that is the improvement of lighting in very specific locations which are the scenes of repeated crime generally reduce crime.

More general increases in street lighting seem to have beneficial crime prevention effects, but this outcome is not universal. This division is by epoch and country, with older North American research yielding fewer positive results than more recent UK research. The most recent British research, carried out in Stoke and Dudley, was impeccable in the methods used and analysis undertaken. It provides firm evidence about the positive effect that improved street lighting can have on crime.

Even untargeted improvements to street lighting generally make residents less fearful of crime and more confident of their own safety at night.

Perhaps unexpectedly, recent research (and reanalysis of data from earlier research projects) shows that street lighting improvements are associated with crime reductions during the daytime as well as during the hours of darkness.[3] This result is of fundamental importance. It means that the beneficial effects of improved street lighting diffuse to periods when the lights are not on, and thus point to the possibility that the benefits are general and not merely restricted to the ability to see potential offenders at night. The most plausible reasons for this pattern are changes in street use, enhanced community pride and an improved sense of area ownership.

A recent meta-analysis of eight studies in North America found that in four studies where night- and daytime crimes were measured, there was a significant reduction in all crime in the relit areas compared to the control areas. Moreover, in a fifth study, there was a highly significant reduction in crimes of violence. Overall, the eight studies taken together showed that improved street lighting reduced crime by 7 per cent. The authors conclude: 'Why the studies produced different results was not obvious, although there was a tendency for "effective" studies to measure both day-time and night-time crimes and for "ineffective" studies to measure only night-time crimes.'[4]

The British study most often cited as showing the absence of positive effects on improved street lighting on crime was published by the Home Office. Although the authors made it very clear that the improvements to street lighting had reduced the fear of crime, they were less convinced that they had discovered a positive effect on crime itself. Some have there-after erroneously assumed that the Home Office is sceptical of the ben-efit of improved street lighting. However, the data from that study in Wandsworth has now been reanalysed and markedly different conclusions are suggested. The original study reached its central conclusion by using changes in daytime crime rates as the benchmark for assessing changes in the number of crimes at night. Data reanalysis shows that if the original study had compared the relit area with the remainder of the police force area in which it was located, a significant positive effect of street light-ing would have emerged, during both the daytime and night-time. In sum, improved street lighting had reduced both crime and the fear of crime.[5]

Reanalysis of data from other earlier studies suggests that street lighting effects are greater in chronically victimised areas, which is of particular importance for the integration of improvements to street lighting in other schemes devised under the provisions of Crime and Disorder Act 1998.

Most recently, the Home Office has published the results of systematic meta-analysis of studies of the effects of both street lighting[6] and Closed Circuit Television (CCTV)[7] on crime. The same two distinguished crimi-nologists conducted both reviews. Of the 18 evaluations of CCTV, they conclude that CCTV had a significant desirable effect on crime, although the overall reduction in crime was a rather small 4 per cent.[8] In comparison, a meta-analysis of street lighting studies found an overall reduction in crime of 20 per cent. In short, street lighting is four times more effective than CCTV in achieving reductions in crime. It may be the case that improved street lighting increases local community pride and informal social control which in turn contributes to a decline in crime rates, rather than directly increasing deterrence.

It may be helpful to summarise the actual conclusions reached by the two independent reviews:

The street lighting study concluded that

- Where street lighting had been improved, there had been an overall reduction in recorded crime of 20 per cent, across all the experimental areas. In the British studies, there was a 30 per cent decrease in crime.
- Improvements in street lighting offer a cost-effective crime reduction measure.

The CCTV study concluded that

- Where CCTV had been installed, there had been an overall reduction in recorded crime of 4 per cent across all the experimental areas.

It was found that CCTV had no effect on violent crimes but had a significant desirable effect on vehicle crimes. When good lighting, CCTV and other security measures were installed in car parks, crime was reduced by 41 per cent.

2.2.3 Improved street lighting and crime prevention: Are there alternatives?

There is no such thing as an all-purpose crime prevention measure. Recognition of this point is crucial. It is widely recognised in the research community that successful crime prevention schemes typically involve integrated packages of measures which combine to produce the desired effect. This does not mean to say that improved street lighting, when introduced as a stand-alone measure, is incapable of having an effect (although it should be realised that some competing crime prevention measures, such as open-street CCTV, need good lighting to work at all at night). Nevertheless, there is some slight evidence that lighting improvements may be particularly valuable in combination with other measures, such as with increases in police patrols, with commercial security surveys and with the rearrangement of available space.

It is not so frequently realised that money for crime prevention is not infinite and that hard financial choices have to be made. Concern with effectiveness of crime prevention has now been overtaken by an interest in cost-effectiveness. It is here, perhaps, that the attractiveness of improvements to street lighting both as part of a crime prevention package and as a stand-alone measure has yet to realise its full potential. On present evidence, improved street lighting is an extremely cost-effective measure. In the Stoke on Trent study, cost–benefit analysis found that within 12 months of street lighting improvements, 695 crimes were prevented. Using Home Office calculations on costs of crime, it was estimated that £400 000 had been saved in crime reduction. Thus for every £1 spent on lighting, £5 was achieved in crime reduction in 1 year. Spreading the capital cost over the 20-year life span of the lighting, the cost–benefit ratio is 1:51. Similarly, in Dudley, for every £1 spent on lighting, £9.40 was saved on crime reduction in 12 months. Calculating the cost–benefit ratio over a 20-year period produced a staggering cost–benefit of 1:112. In 12 months, 641 crimes had been prevented and £500 000 had been saved in crime reduction.[9]

When cost-effectiveness is considered, the situation becomes more complicated. No research is to hand that demonstrates that, its popularity aside, CCTV reduces either crime or the fear of crime cost-effectively, although it is of undeniable benefit to occasionally allow the police to apprehend offenders in the act. Nevertheless, when compared to improved street lighting, CCTV is expensive to install, involves considerable additional running costs, and relatively frequent and large investment in equipment replacement.

The introduction of cost-effectiveness concerns into comparative crime prevention evaluation has once again elevated improvements to street lighting to the forefront of the agenda. The concern is now to pinpoint the precise circumstances under which improved street lighting might reduce crime, in addition to the fear of crime. It is contended that the aim should now be to use context-appropriate lighting schemes as part of a full repertoire of crime reduction tactics. This point is developed below.

2.2.4   The role of lighting in the development of crime control strategies

There are many diverse ways of preventing crime, of which improved street lighting is one. What follows are evidence-based policy recommendations about the role of lighting in the development of crime control strategies.

- The presumption must now be that improved street lighting be considered as one element in local strategies under the Crime and Disorder Act 1998, alongside other physical and social improvements.

- The evidence that the effect of improved street lighting is greatest in the most crime-prone areas is of crucial importance for the Government's concerns with reducing inequalities in the distribution of crime and disorder. Street lighting enhancements seem to have their greatest effects where one would most want them to.

- Improving street lighting, when it reduces crime, tends to reduce daytime crime as well as crime at night. This certainly means that a simple view of street lighting as having its effects through increasing surveillance is not tenable. It probably means that street lighting improvements which are accompanied by other measures to consult and inform residents, and otherwise enhance social cohesion, will stimulate the mixture of 'active ingredients' underlying positive street lighting effects. Such considerations should be incorporated in local crime reduction strategies under the 1998 Act.

- Crime and disorder is distributed very unequally. For example, in 1991, some 60 per cent of assaultive crime occurred in the 10 per cent most crime-prone areas. This means that crime prevention measures across wide areas with varying crime problems will generally not be regarded as the most cost-effective use of resources. In the CCTV challenge process, local bids had to justify locations for camera installation. In the Government's Crime Reduction Programme, the Local Initiatives element will unquestionably be subject to a similar process of assessment for funding. The criteria of such assessment will include originality and focussed effort. For this reason, schemes which involve lighting small areas or the innovative use of lighting will have an advantage over those which involve lighting of large areas and those in which the mechanism of operation is explored. It is acknowledged that street lighting has economics of scale and that its installation is only incidentally crime preventive. Thus there will be a tension between the extent and nature of street lighting desirable for other purposes and that for crime reduction.

- Section 17 of the Crime and Disorder Act 1998 requires all local authorities, including joint authorities and police authorities, to consider crime and disorder reductions while exercising all their duties. This has implications for all local authority functions, but the particular case of street lighting will be developed here. This means, presumably, that any decisions on lighting (or positive decisions not to relight) locations will have to have crime and disorder implications factored in. The merit of Section 17 is that it recognises that policy decisions, outside policing and criminal justice, drive rates of crime. It may substantially change the perspective of authorities addressing lighting and all other responsibilities.

- The unhappy consequences of dispute about whether or not street lighting affects crime have been that the mechanisms involved have been neglected. Notions of differential lighting to shape use of space in crime reductive ways (perhaps particularly of 'youths causing annoyance') have never been addressed by research or local innovation. In particular, the contrast between private security lighting, where illumination is triggered by a sensor and lighting in public spaces, which is almost always time triggered, is remarkable. A debate is necessary about this contrast, and about expanding ways of thinking about illumination and its role in crime reduction.

2.3 Light pollution

2.3.1 Adverse effects of outdoor lighting

All living things adjust their behaviour according to natural light. The invention of artificial light has done much to safeguard and enhance our night-time environment but, if not properly controlled, obtrusive light (commonly referred to as light pollution) can present serious physiological and ecological problems.

Light pollution, whether it keeps you awake through a bedroom window or impedes your view of the night sky, is a form of pollution and could be substantially reduced without detriment to the lighting task.

Sky glow, the brightening of the night sky above our towns and cities; glare, the uncomfortable brightness of a light source when viewed against a dark background; and light trespass, the spilling of light beyond the boundary of the property on which the light source is located, are all forms of obtrusive light (Figure 2.1). This is not only a nuisance, it wastes electricity and thereby large sums of money, but more importantly it helps destroy the Earth's finite energy resources, resulting in the unnecessary emissions of greenhouse gases.

For any outdoor lighting installation, its lighting impact will not be the only environmental impact and may not even be the most important. Other factors relating to the uses which are facilitated by the lighting system could, and often are, more significant than the lighting system itself, for example noise, traffic or parking.

However, lighting is often the focus of complaints because it is, by nature, highly visible and is the means by which the conduct of the night activity

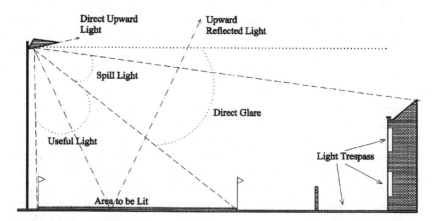

Figure 2.1 Light pollution terminology.

is made possible. The potential effects of the lighting should therefore be assessed, as part of the overall impacts of a development, by the relevant development approval authority.

The impact of a lighting installation on the environment is not limited to the imposition of obtrusive light. The designers of a lighting installation should utilise luminaires and light sources that efficiently direct the light only into the area required, thereby minimising also energy consumption and use.

2.3.2 Influence of surrounding environment

The perception of a lighting installation will be significantly influenced by its environment. Therefore a series of 'Environmental Zones' have been agreed upon which have been recommended for adoption by Local Planning Authorities in order for them to better control potential light pollution. These are shown in Table 2.1.

In relation to Table 2.1, the following factors should also be noted:

- There is usually a greater potential for complaints where the lit area is surrounded by residential development.
- The topography of the area surrounding the lighting installation can also be important. Areas that are at a lower level than that of the lighting installation should be particularly considered, where a direct view of the luminaires is possible.
- Physical features such as adjacent tall buildings, trees and spectator stands can help in restricting light spill beyond the boundaries of the development.
- The presence or absence of other lighting in the immediate area and the type of lighting involved. The effect of the proposed lighting will

Table 2.1 Environmental zone classification

Zone	Category	Examples
E1	Intrinsically dark areas	National Parks, Areas of outstanding natural beauty, etc.
E2	Low district brightness areas	Rural or small village locations
E3	Medium district brightness areas	Small town centres or urban locations
E4	High district brightness areas	Town/city centres with high levels of night-time activity

Note
Where an area to be lit lies on the boundary of two zones or can be observed from another zone, the obtrusive light limitation values used should be those applicable to the most rigorous zone.

be lessened where the surrounding area is reasonably well lit, for example arterial road lighting or lighting from adjacent commercial developments.

- Other aspects that may effect how the lighting installation is perceived within the environment are

  (i) areas of special significance, for example areas having cultural, historical or scientific importance;

  (ii) harbours, airports, waterways, roads or railway systems where spill light from the proposed development may interfere with the visibility of signalling systems; or

  (iii) community and scientific optical observatories where upward light from the proposed development because of resulting sky glow may interfere with astronomical observations.

2.3.3   Relevant lighting parameters

2.3.3.1   Effects on the natural environment

Effects on the natural environment, which includes sight of the stars at night, is emphasised by the brightening of the sky caused by the scattering of light from an installation into the atmosphere, producing the luminous glow that we term 'sky glow'.

Sky glow is caused by both reflected light and direct light from an installation, and therefore restricting design illuminances to the minimum necessary for the application will provide important mitigation. However, the most effective action is to control the amount of direct light emanating from close to and above the horizontal, that is, by limiting the upward light output ratio of the luminaire. Unfortunately, this will only be relevant if the luminaire is mounted horizontally in a fixed position, as in road lighting. To cater for those situations where the luminaire may be at any angle, the parameter 'upward light ratio' (ULR) is more appropriate.

2.3.3.2   Effects on residents

The effects on residents generally involves a perceived change in amenity arising from either of the following:

- The illumination from spill light being obtrusive, particularly where the light enters rooms of dwellings which are normally dark, for example bedrooms. The illuminance on surfaces, particularly vertical surfaces, is an indicator of this effect.

- The direct view of bright luminaires from normal viewing directions causing annoyance, distraction or even discomfort. The luminance of a luminaire, in a nominated direction, is an indicator of this effect. However, because luminance data is not normally provided by luminaire manufacturers and because of difficulties associated with the measurement of luminance, recommendations are normally expressed in terms of the luminous intensity in specified directions.

(Note: Values of these parameters that are acceptable during the earlier hours of the evening may become intolerable if they persist at later times when residents wish to sleep. It is therefore usual to give values of before and after 'curfew'; a time after which more stringent controls should be applied.)

2.3.3.3 Effects on sightseers

The effects on sightseers of over-bright or unsuitably coloured decorative lighting and signage will be to regard the overall lighting effects as obtrusive rather than enhancing the night-time scene. The relevant indicator will be the luminance of the surfaces. The acceptable luminance of signs will depend on the size of the surface viewed (see ILE Technical Report No. 5).

2.3.4 Recommended limits for lighting parameters

Table 2.2 is reproduced from the ILE Publication 'Guidance Notes for the Reduction of Light Pollution – 2000' which is widely regarded as the definitive document on this subject. The values have been agreed amongst an international group of lighting professionals as well as astronomers, planners and environmentalists.

2.3.5 Design guidelines

- *Examination of alternatives* – Before arriving at the final design, consideration should be given to alternative lighting systems with respect to their capability of fulfilling both the functional and environmental design.
- *Location of illuminated area/activity* – When there is some flexibility as to where the illuminated area/activity can be placed, it should be located and oriented where it will have the least effect on existing or potential developments. Advantage should be taken of any screening which may be provided by the surrounding topography or other physical features, for example buildings, trees or earth embankments.

Table 2.2 Obtrusive light limitations for exterior lighting installations

Environmental zone	Sky glow: Maximum ULR for the installation (%)	Light into windows: Maximum vertical point illuminance (in lux) not greater than[a]		Maximum light source intensity *I* (in kcd)[b]		Building luminance before curfew[c]	
		Before curfew	After curfew	Before curfew	After curfew	Maximum average luminance (in cd m^{-2}) not more than	Maximum point luminance (in cd m^{-2}) not more than
E1	0	2	1*	0	0	0	0
E2	2.5	5	1	20	0.5	5	10
E3	5.0	10	2	30	1.0	10	60
E4	15.0	25	5	30	2.5	25	150

Notes

a Light into windows – These values are suggested maxima and need to take account of existing light trespass at the point of measurement. * Acceptable from public road lighting installations ONLY.

b Source intensity – This applies to each source in the potentially obtrusive direction, outside the area being lit. The figures given are for general guidance only and for some large sports lighting applications, with limited mounting heights, may be difficult to achieve. If the aforementioned recommendations are followed then it should be possible to further lower these figures.

c Building luminance – This should be limited to avoid over lighting, and relate to the general district brightness. In this reference building luminance is applicable to buildings directly illuminated as a night-time feature as against the illumination of a building caused by spill light from adjacent floodlights or floodlights fixed to the building but used to light an adjacent area.

- *Selection of luminaires* – The luminaires selected should have an appropriate light output distribution for the application and, when correctly located and aimed in accordance with the design, should not emit excessive light outside the boundaries of the property on which the installation is sited. To facilitate an adequate assessment of compliance with the relevant criteria, photometric performance data will be required for all angles at which light is emitted by the luminaires, not just for the angles which define the useful light output, for example the beam of a floodlight.

  For environmentally sensitive sites, particular attention should be given to the selection of luminaires with good spill light control. Systems for classifying floodlights in terms of their beam shape and divergence should be used for the specification of design requirements best suited for the particular application. Louvres, baffles or shields may be fitted to floodlights to control spill light although account will need to be taken of the effect of the devices on the performance of the lighting system.

- *Siting and aiming of floodlights* – The required locations for floodlights are often determined by the nature of the activity for which the lighting is provided. However, small departures from the recommended positions may be acceptable if this will result in a greater degree of control of the spill light.

As illuminance generally reduces in proportion to the inverse of the square of the distance from the floodlight, it follows that the greater distance the floodlights can be set back from a property line, the lower will be the illuminances at and beyond that property line. The objective of the design should be to ensure that, as far as is practicable, direct view of the bright parts of the floodlights is prevented from positions of importance at eye-height on neighbouring properties. Where possible, advantage should be taken of the shielding that may be provided by trees, earth embankments, spectator stands or other existing physical features.

When determining the mounting height of the luminaires, consideration should be given to the following:

- Higher mounting heights can often be more effective in controlling spill light because floodlights with a more controlled light distribution (i.e. narrower beam) may be used, and the floodlights may be aimed in a more downward direction, making it easier to confine the light to the design area.
- Lower mounting heights may have the advantage of making the lighting installation less obtrusive by day, but can accentuate its effect on the night environment by increasing the light spill beyond the property boundaries because, to illuminate the space satisfactorily, it will often be necessary to use floodlights with a broader beam and to aim the floodlights in directions closer to the horizontal than would normally apply.

2.3.6 Methods of mitigation

As noted previously, one of the best ways of mitigating the effects of a lighting installation is to only light what needs to be lit, and to the minimum level required for the task. The quality of the lighting is not necessarily improved by adopting higher lighting levels. Improved visual conditions for the participants of the activity being lit can often be obtained by giving greater attention to the uniformity of illuminance over the design area and the control of glare from luminaires.

The use of well-designed, good quality equipment is also an advantage, which should be available with a wide selection of accessories as indicated below.

2.3.7 Illustrations of luminaire accessories for limiting obtrusive light

Figure 2.2 Internal baffle.

Note: This luminaire also has an external protective grill.

Figure 2.3 Cowl (or hood).

Figure 2.4 'Barn doors' shield.

(a)

Figure 2.5 External louvres: (a) circular and (b) straight.

(b)

Figure 2.5 (Continued).

2.4 Strategic urban lighting plans

2.4.1 Introduction

Since 1989, urban lighting strategies (or strategic urban lighting master plans) have been a major feature of the 'City Beautification' process in the UK as well as in other European, North American and Asian cities. In the UK, more than 26 such strategic plans have been commissioned for a range of urban centres, starting with the influential Edinburgh Lighting Vision in 1989/90. Major UK towns and cities, such as Leeds, Coventry, Chester, Croydon, Cambridge and Liverpool, have looked to urban lighting strategies to lay out a rational, comprehensive plan of their future lighting needs and as a marketing tool, to help raise funds for the implementation of the plan.

2.4.2 Origins of the urban lighting plan

The concept of the coordinated, strategic urban lighting plan, encompassing all types of lighting within a defined urban area, emerged in the late 1980s in three specific cities in three different countries: Milwaukee in the USA; Lyon in France; and Edinburgh in Scotland.

In Edinburgh, the Edinburgh Lighting Vision Plan was perceived in 1989 by the Scottish Development Agency (SDA) as a way of stimulating and developing the Scottish capital's tourist economy outside the peak summer months. The main motive for the plan was the negative experience of the city of Glasgow in the mid-1980s, which had invested substantial sums of money in a miscellaneous, uncoordinated series of lighting initiatives, chosen in an arbitrary manner, which seemed to have very little net visual effect on the night-time lit appearance of the city. However, Glasgow are now implementing an updated plan.

The SDA decided that a coherent plan for lighting Edinburgh was required, which could guide investment policy in a more rational and fruitful manner over the space of several years, according to the social, development and economic needs of the city as a whole. Another motive was the growing concern about 'light pollution' and the unplanned effects of exterior and security lighting, raised by astronomers, environmentalists, planners and lighting designers themselves. It was felt that a planned approach to urban lighting would also help combat this growing problem.

2.4.3 Lighting strategy methodology

The method of researching the Edinburgh Lighting Vision Plan was to start with a thorough analysis of the central area of the city within the main bypass road. They used photographs and plans (Figure 2.6) to determine how the city 'revealed' itself to the visitor, as a guide to future lighting priorities. The main features of this detailed analysis that follows remain the basis of the methodology:

- The principal 'gateways' into the central city area, defined as the major traffic 'nodes' on the approach routes, at which the city itself became visible
- The main 'skyline' architectural features of the city
- The main topographical features of the city – hills, valleys and so on
- The main vistas and long, medium and short views of these features
- The main historical/architectural nature of different 'zones' within the city
- The main patterns of night-time use by residents, visitors and tourists
- The existing lighting.

Flowing from this analysis it is possible to develop a specific set of lighting proposals for implementation over a period of, say, 5–10 years. The main features to consider are

- New, highly visible lighting treatments of key 'gateway' features on the main approach routes, to create a genuine sense of 'arrival' within the city
- Selective illumination of key 'skyline' architectural features, visible from large parts of the city (in Edinburgh, the Castle and monuments on Carlton Hill)

Figure 2.6 First sketch for the Edinburgh Lighting Vision Plan.

- A coordinated change to the colour temperature of the road/street lighting in particular areas to complement the historical/architectural character of that part of the city (in Edinburgh a 'cool' treatment for the geometrically laid out streets of the Georgian 'New Town' and a 'warmer' treatment for the medieval Old Town, to echo the light sources of that period, i.e. candles and torches). The design of the lighting equipment itself would also accentuate this architectural differentiation
- Maximum luminance recommendations for, say, three categories of city zone, to keep their relative brightness in balance – and to prevent competitive 'light wars' between commercial premises
- A series of lighting investment priorities, to make the maximum visual impact with the funding available
- The appointment of a City Lighting Manager to oversee and popularise the Plan, who would also work to educate building owners and local politicians on its recommendations and implementation requirements
- In addition, an invaluable part of almost every lighting strategy to date has been the mounting of one or more temporary (one or two night) 'site tests' or demonstrations of specific lighting proposals, using portable lighting equipment. This can play a useful role in demonstrating the practicality and attractiveness of the plan to council officers, elected representatives, private-sector partners and the media.

With some differences in emphasis, this methodology and general approach informed most of the 26 full-blown lighting strategies that have been undertaken in the UK since 1989. The Edinburgh Lighting Vision was hugely successful in terms of its broader influence and, in particular, in raising an awareness of the importance of urban lighting in the UK.

2.4.4   The aims of a strategic urban lighting plan

The main aims of a strategic urban lighting plan can be summarised as follows:

- The provision of a long-term plan for the development of all forms of lighting over 5–10 years and beyond time span
- The visual unification of the disparate night-time components of urban centres, leading to the creation of a coherent night-time identity
- The accentuation of a town/city's main 'gateways' or entry-points, to create a genuine sense of 'arrival'
- The after-dark presentation of a city's architectural and heritage assets to best effect
- The resolution of conflicts between the different lighting needs of various types of users – pedestrians, workers, tourists, drivers and so on
- The minimisation of light pollution, light trespass and visual imbalances in the night-time scene
- The re-prioritisation of the visual needs of the night-time pedestrian and the creation of a safer, more pedestrian-friendly night-time environment
- The longer-term goal of stimulating and developing the evening economy, and giving the town/city a competitive edge.

In addition, the strategic lighting plan might itself be used as a marketing tool for raising additional funding for implementing its proposals, from the private sector, local or national government or other sources.

2.4.5   The main factors in the success of lighting strategies

2.4.5.1   General

The effectiveness of plans has been mixed, with some plans, such as Leeds and Coventry, changing the local lighting culture. However, others were virtually stillborn and simply gathered dust on local authority's shelves. And yet others have been compromised by the unwillingness or financial inability of local authority lighting departments to embrace their radical vision.

In addition to the most obvious factor, the quality and appropriateness of the strategy itself, the success or failure of strategic lighting plans depends on a number of inter-related factors:

- The nature, skills and experience of those undertaking the strategy
- The role, responsibility and power of those commissioning the strategy
- The nature of the commissioning local authority and its role in relation to public lighting (i.e. does it control road/street lighting or is that in the hands of another agency, such as the County Council or Highways Agency?)
- The direct involvement or otherwise of the authority's lighting department in the evolution of the plan
- The geographical scope of the plan and the precision of the initial brief
- The level of understanding of the plan within the commercial and business sector and private building owners, and their responsiveness and engagement with it
- The balance of ambition vs practicality embodied in the plan
- The amount of special funding available for short-term implementation of the plan's proposals, or at least a number of exemplary benchmark projects
- The usefulness of the strategy document as a marketing tool and to raise public awareness and additional support within the wider society.

2.4.5.2   Who does them?

In the vast majority of cases, strategic lighting plans in the UK have been undertaken by independent lighting design consultants (sometimes in partnership with an architectural practice). These have either been direct appointments by the commissioners or were subject to a public tendering competition.

In a minority of cases, lighting manufacturing companies have undertaken specific lighting plans. No lighting strategies have been generated directly by local government lighting departments. In general, commissioners seem to have recognised that the application of external, unbiased expertise of independent lighting design consultants was likely to provide them with the kind of objective, comprehensive long-term view necessary to generate an effective and useful plan. Most independent UK lighting consultancies include specialists from lighting engineering, architectural, theatre and product design backgrounds, skills that can work well together in generating long-term wide-ranging lighting strategies. This is not to say that a local government lighting department could not lead a team equipped with the necessary skills.

2.4.5.3   Who commissions them?

Broadly speaking there have been two types of commissioner. In the early days, strategies were often commissioned by various government-funded local Development Agencies or City Centre management bodies or even national government bodies. Examples include Edinburgh (SDA), St Andrew's (Fife Enterprise) and Londonderry (Department of Environment, Northern Ireland). This had important implications.

Although such bodies have been relatively well funded and were often able to implement some key lighting proposals (mainly building and feature lighting) at an early stage, they tended to have poor or non-existent relations with local government officers, particularly street lighting departments. This usually meant that few or none of the proposals for integrating street lighting into the plan, through a process of gradual modification of light sources or luminaire type, were likely to be carried out. The Edinburgh Lighting Vision plan, commissioned by the SDA, is a good example of this.

Later in the early 1990s, as Development Agencies were phased out by central government, local government authorities themselves began to recognise the importance of lighting strategies, as part of their regeneration plans. However, in most cases, the primary local government commissioners of lighting plans tended to be tourism departments, economic development agencies, regeneration or conservation officers or, later still, local planning departments.

Once again, in many cases, local authority street lighting engineers were only peripherally involved or not involved at all in either preparing the brief or commissioning the plan. One reason for this is the strong tendency towards autonomous, departmentalised thinking which has marred UK local government until recently and which has resulted in a minimum of cross-departmental co-operation. In the few cases where a sympathetic local authority chief lighting engineer was involved in the commissioning and briefing process, such as in Leeds in 1992, the effects were significant. The lighting strategy proposals became firmly rooted in the local culture and succeeded in generating a number of long-term changes, in the lighting of both roads/streets and private and local authority buildings. Leeds also exemplified another interesting and useful model for winning broader support for the plan, the formation of an 'arms-length' partnership body, involving the local authority and other outside agencies (in the case of Leeds the Leeds Development Corporation) to commission, administer and promote the lighting strategy.

Most recently, local planning authorities have begun to work much more closely with local lighting engineering departments in the elaboration of the plans. This is exemplified by the recent plans for Cambridge and Coventry.

2.4.5.4   The nature of the local authority and its responsibility for lighting

In the context of the UK the type of local authority that is involved is a very important factor too. It is a peculiarity of the structure of local government in the UK that various aspects of public lighting can fall under the remit of different agencies, mainly either the County Council or the Highways Agency. Only in larger metropolitan unitary authorities is the local council responsible for all aspects of public lighting, amenity lighting, the lighting of secondary streets and the lighting of major highways. In such cases (and Leeds is a very good example) the authority can take on board proposals for changing the colour temperature of street and highway lighting, and (funds permitting) implement them without reference to other bodies. This makes the achievement of a comprehensive lighting plan so much easier.

In many other cases, with smaller non-unitary authorities, such as Chester, Scarborough and Cambridge, there is a divided responsibility for public lighting. The local authority will be responsible for amenity lighting and sometimes the illumination of side streets, with the County Council or Highways Agency taking on all the lighting of major highways and through routes.

In Scarborough, the consultants proposed the phasing out of low pressure sodium (SOX) lighting in residential areas. The local authority, which had commissioned the Strategy, was very much in favour of the proposal, but had no power to persuade the County Council to implement the change. Although a number of lighting schemes of key historic buildings in Scarborough, flowing from the Strategy, are currently under implementation, they look set to remain beacons of exemplary architectural lighting in a sea of orange road lighting. It is this peculiar anomaly of UK local government structures that remains the major barrier to the adoption and implementation of a comprehensive lighting strategy in many areas of the UK.

2.4.5.5   The geographical scope of the strategy

It is very important that the scope of the Lighting Plan or Strategy has a logical and realistic geographical scope; otherwise its proposals will be diluted or rendered totally unrealisable. Most lighting plans tend to address the lighting of the historic or commercial heart of towns or cities and usually encompass a strictly delineated area, bounded by the ring road or other natural boundary. Some plans, such as Blackburn, Hamilton or the Stratford in East London, have addressed a much smaller part of the total town/city area. While this makes implementation in a short period more likely, there is the danger that the original, unreformed lighting of adjacent areas will threaten to overwhelm any improvements.

The original Edinburgh Lighting Vision plan covered an area larger than the main tourist area of the city centre, extending out to the major routes into the city through the suburbs. It is very unlikely that the proposals for this wider area outside the city centre will ever be realised. In the worst case LDP's lighting strategy for Lanarkshire in Scotland, commissioned by Lanarkshire Development Agency in 1995, attempted to elaborate a lighting plan for an entire county covering 2000 sq km, on a 25-year timescale. The value of the implementation work would have run into several millions of pounds. As a result, this over-ambitious plan was virtually stillborn and little ever came out of it, despite optimistic projections of its economic benefits by business consultants.

2.4.5.6 Vision versus practicality

Successful lighting plans must obviously address the real visual and lighting needs of cities or towns, and must embody practicable and realisable lighting proposals that will work in the short-to-medium term. At the same time, they must embody an element of the visionary, the unexpected, the unorthodox, to enthuse and excite the broader community, private-sector partners and so on.

Two differing perceptions have emerged. One is that:

> Lighting strategies have become more practical in recent years. They have been taken more into the heart of local authorities. They have become more pedestrian [in both senses of the term] and mundane, more mainstream, more like any other area of urban planning, like paving. In fact they have become less a vision, more a planning exercise.[10]

The contrasting view is:

> The major additions to lighting strategies have been the moves to the '24 Hour Economy'. People who use urban space now expect so much more – the city is becoming an entertainment and leisure venue and we have to use more theatrical and dramatic lighting techniques, but in a controlled manner. It's another layer of lighting over and above the base lighting for pedestrians and traffic.[11]

Whilst the former view is basically true, the aim must be to ensure the latter is predominant. Figure 2.7 shows Full-scale trials at Croydon.

2.4.6 The effectiveness of lighting strategies

Of the 27 lighting strategies listed below, only five (Newark, Lanarkshire, Isle of Thanet, Ashford and Nottingham) have resulted in very few actual

Figure 2.7 Full-scale trials at Croydon.

lighting outcomes. Four (Hull Transport Corridor, Belfast, Liverpool and Hamilton) have been completed too recently to assess in terms of their effectiveness. In four cases (Guildford, Thornton Heath/South Norwood, Blackburn and Three Towns, Isle of Man) the outcomes are unknown. Of the remaining 14 lighting strategies, nine (Edinburgh, Leeds, Leeds Riverside, Londonderry, Stratford in East London, Coventry, Croydon, Scarborough and Portsmouth and Gosport Harbour) have seen considerable implementation and can be regarded as a broad success. St Andrew's, Chester, Pool of London and Cambridge have seen some implementation.

2.4.7 Conclusions

One conclusion that cannot be avoided is that the broader success of Lighting Plans depends on the force and enthusiasm with which lighting plans are implemented and the 'carrots and sticks' that can be brought to bear on private building owners in particular to abide by the broad precepts of such Plans. But as long as they remain purely advisory or pure propaganda tools, their influence will remain limited. In the longer term many lighting practitioners believe (though this is by no means universal) that their ultimate adoption and realisation will be dependent on their having some statutory force, through their incorporation into wider planning law. This will demand new national legislation and will be subject to a much wider debate.

In the meantime, some of the main pre-conditions for a successful Strategic Lighting Plan remain:

- The manageable and logical geographical scope of the Plan
- A set of realisable objectives and proposals, combined with an exciting 'visionary' element to catch the public imagination
- Popularisation of the visual and economic benefits of the Plan to the wider public
- Integration of lighting with other forms of urban regeneration or improvement
- Establishment of a public–private partnership to administer and popularise the Plan
- Adequate sources of funding to finance the early stages of implementation – particularly exemplary 'benchmark' projects
- The existence of a unitary authority with full responsibility for all aspects of road and amenity lighting
- The achievement of a critical mass of new lighting, at an early stage, to persuade the public and local media and to cement the various funding partners
- The participation of co-operative local lighting engineers and highways authorities in the Plan's brief and elaboration.

2.5 Health and Safety

2.5.1 Introduction

There is a large amount of legislation and guidance associated with Health and Safety and the volume of such information can often present a bewildering picture. However, the purpose of Health and Safety legislation is to prevent injury, accidents or ill health, first by taking positive measures which will have the greatest impact on the largest number of people and then taking further more specific measures. These measures are set in general terms as aims and objectives within the relevant legislation and the detailed measures should be ascertained using competent persons and through consultation. By adopting this relatively straightforward approach to the management of Health and Safety, it is possible to define strategies to deal with this area.

This section is not intended to give a complete synopsis of all the Health and Safety information which might be applicable within the Outdoor Lighting Environment, rather the intention is to define the principles governing the management of health and safety and to deal with a few specific items of legislation which are particularly relevant. Further information can be obtained direct from the Health and Safety Executive on both legislation and guidance to that legislation.

2.5.2   The body of law

Before we look at some of the legislation, it is as well to have a basic understanding of law. There are two main branches of law – criminal and civil – and together these make up the body of law. In any organisation it is necessary to conduct oneself in such a way as to avoid both criminal and civil actions as either of these will inevitably cost the organisation both time and money as well as affecting other areas such as reputation.

Criminal law concerns itself with offences against the State and in general relates to breaches of Acts or Regulations made under the authority of Parliament. Such breaches only need to occur for a criminal act to have taken place, that is there is no requirement for any injury or loss to have occurred as well. The action of a criminal court is to establish the facts beyond reasonable doubt and if the accused is guilty, to punish the accused in the hope of deterring future criminal action.

Civil law concerns itself with compensation or some other action between individuals, and in general some loss or injury needs to be established, on the balance of probabilities, which is a lesser standard than that of the criminal case. The purpose of civil law is to remedy an existing situation to the benefit of the 'injured' party bringing the action. Compensation payments can be, and have been, substantial sums of money.

It should be borne in mind that the history of law has shown a mirroring of civil law by subsequent criminal law and that by satisfying Acts or Regulations, it is normally the case that one can prevent or limit potential civil claims.

2.5.3   Duty of Care

One of the key principles underlying both civil and criminal laws is that of the 'Duty of Care'. This can be either an implied or express duty, and is a duty to take care not to harm an individual or the property of an individual.

In order for a duty of care to apply, there has to be a 'relationship' between the two parties. This has been established in case law such that a duty of care is owed to one's 'neighbour', that is anyone who might reasonably be considered could be affected by the acts or omissions of the other party. In addition, the duty of care has been expanded, again through case law, to take account of individuals and their particular circumstances and such a duty is often called a Duty of Common Humanity. This means that the standard of the duty of care owed to a child, for example, is higher than that owed to an adult – as the child has less knowledge and experience and therefore greater care needs to be taken. This principle has been reflected within, for example, the Management of Health and Safety at Work Regulations where it deals with young persons.

2.5.4 Qualified and absolute duties

Most of the Acts or Regulations we shall be considering in this section impose qualified duties on persons – that is the duty must be carried out as far as is reasonably practicable. However, there are a number of instances, for example within the Electricity at Work Regulations, where the requirements are absolute.

An absolute requirement is one which must be met regardless of cost or any other consideration. A duty which is 'reasonably practicable' is one in which the quantum of risk on the one hand must be balanced against the time, cost, trouble and difficulty of removing that risk on the other hand, and a judgement must be made accordingly. However, where the risk is great (e.g. that involving death by electrocution) and the time, cost, trouble and difficulty of removing that risk is small, then it can be argued that the duty approaches that of an absolute duty.

2.5.5 Health and Safety hierarchy

It is useful to understand the hierarchy of legislation and corresponding advice to ensure that we weight the information available accordingly. Legislation, of course, must be met as the law of the land. Approved Codes of Practice (ACoPs) published by the Health and Safety Commission have the effect that a breach of such an Approved Code is not, in itself, a breach of the legislation. However, by demonstrating that one is following an ACoP, one can show compliance with the legislation. Within the hierarchy of information, below legislation and then ACoPs, there then follows HSE Guidance Notes, European/British Standards and then industry codes of practice.

It should be noted that as Health and Safety legislation has become less prescriptive and more about setting policies and objectives, ACoPs and other standards and guidance have become more important in assisting individuals and organisations to fulfil their duties.

2.5.6 The Health and Safety at Work etc. Act

The Health and Safety at Work etc. Act (HASAWA) is the over-arching legislation within and under which all other health and safety legislation is derived. The HASAWA is an enabling Act in that under Section 15, the Secretary of State can direct the Health and Safety Executive and Commission to draw up Regulations without requiring the full process of Parliament to be implemented.

The HASAWA essentially covers three main areas for the practical implementation of health and safety – the duties of employers to employees, the duties of employers to non-employees and the duties of employees. There are other areas to the Act covering issues such as administration of health and safety matters, enforcement, etc. and in particular Section 40 which 'reverses' the usual application of law in placing the burden of proof on the accused.

Section 2 covers the duties of employers to employees and requires employers to 'ensure, as far as is reasonably practicable, the health, safety and welfare at work of all his employees' and in particular to ensure, as far as is reasonably practicable:

- Safe plant and systems of work
- Safe use, handling, storage and transportation of articles and substances
- The provision of appropriate information, instruction, training and supervision
- Safe place of work including access and egress
- Safe working environment.

Section 3, which is particularly relevant when looking at outdoor lighting, requires employers to conduct their undertaking in such a way that persons other than employees (e.g. this could include the general public) are not exposed to risks to their health and safety.

Section 7 covers employees' duties and requires them to 'take reasonable care for the health and safety of himself and of other persons who may be affected by his acts or omissions' and to co-operate with the employer. In addition, under Section 8 it is forbidden for any person (i.e. this is not limited to employees) to interfere with or misuse anything provided by the employer to meet their obligations under health and safety legislation.

2.5.7 The Management of Health and Safety at Work Regulations

The Management of Health and Safety at Work Regulations (MHSWR) expands and amplifies the HASAWA and together with that Act forms the framework for all other Health and Safety legislations. In particular within the MHSWR there is:

- An absolute requirement to carry out suitable and sufficient risk assessments
- A requirement to establish and give effect to procedures to be followed in the event of serious and imminent danger
- A requirement to take additional steps to protect the health and safety of young persons (i.e. those under 18 years old) and new or expectant mothers
- A requirement for employers to appoint one or more competent persons to assist him.

By comparing the two sets of legislation one can find examples of areas where the MHSWR amplifies the requirements of the HASAWA such as:

- HASAWA Section 2(2) requires safe plant and systems of work; MHSWR Regulation 3 expands this by placing a requirement for risk assessments to be carried out
- HASAWA Section 2(2) requires the provision of information, instruction, training and supervision; MHSWR Regulations 10 and 13 expand this by requiring information to be provided on risk assessments and for training to be carried out at defined times (e.g. on induction, where new technology is being introduced, etc.)
- HASAWA Section 7 covers employees' duties; MHSWR Regulation 14 requires employees, amongst other things, to report dangerous situations and any shortcomings in the protection arrangement for health and safety.

One of the most important areas identified by the MHSWR is the requirement to assess the risks to those who might be affected by the undertaking and to determine what measures should be taken to comply with the relevant legislation or 'relevant statutory provisions'. Within the guidance produced by the Health and Safety Executive, there is specific guidance on how to carry out risk assessments and it is now generally accepted that the steps to producing a risk assessment are

- Identify the Hazards (a Hazard is anything with the potential to cause harm)
- Identify who might be harmed
- Identify the existing controls, if any, in place
- Identify the standard to be reached (i.e. by reference to the relevant legislation, ACoPs and other guidance)
- Identify what actions need to be carried out, by whom and by when in order to reach the standard
- Record the findings
- Review the risk assessment and revise it, if necessary.

In applying preventative or protective measures, the employer should adopt the following hierarchy for dealing with risks:

- Avoid or eliminate risks
- Evaluate the remaining risks and combat these at source
- Adapt the work to the individual
- Take account of new technology

- Replace the dangerous by the less dangerous (or preferably the non-dangerous); isolating the employee from the risk; controlling the risk by working arrangements, protective equipment and so on as part of a coherent risk prevention policy
- Give priority to those measures which will positively affect the greatest number of people and provide the greatest benefit (i.e. collective measures over individual measures)
- Provide information, instruction and training so that everyone who might be affected understands what they need to do.

Risk Assessments are usually, to a greater or lesser degree, generic in that in order for the assessment to be anything other than generic it needs to be carried out for each individual taking into account that individual's knowledge, experience, education, aptitude, psychological makeup, etc. However, by adopting the principles set out in the MHSWR and associated guidance, and taking into account the requirement to deal with young persons and new or expectant mothers as separate groups, it should be possible to produce 'suitable and sufficient' risk assessments which are both realistic and practical.

2.5.8   The Provision and Use of Work Equipment Regulations

The Provision and Use of Work Equipment Regulations (PUWER) cover any equipment used by an employee at work (e.g. screwdrivers, hammers, ladders, vehicles, photocopiers) and in addition lifting equipment must also comply with the Lifting Operations and Lifting Equipment Regulations (LOLER).

PUWER place requirements on employers or those persons in control of work equipment to ensure that

- Equipment is suitable and only used for operations for which it is suitable
- Equipment is maintained in efficient working order and good repair
- Equipment is inspected at defined times
- Equipment is used only by persons who have received suitable information, instruction and training
- Equipment is provided with appropriate guarding, lighting, markings, warnings and warning devices.

In addition there are further requirements for mobile work equipment to ensure, amongst other things, that employees are not harmed through such equipment becoming unstable or being operated by unauthorised persons.

2.5.9   The Lifting Operations and Lifting Equipment Regulations

The Lifting Operations and Lifting Equipment Regulations (LOLER) applies to all equipment used at work for lifting or lowering loads (e.g. scissors lift, loader crane, Mobile Elevated Work Platform) and includes lifting accessories such as chains and slings. Such equipment must also comply with the requirements of the PUWER.

LOLER place requirements on employers or those persons in control of lifting equipment to ensure that

- Lifting equipment is of adequate strength and stability
- Lifting equipment is positioned and installed to minimise risks
- Lifting equipment is suitably marked to indicate its Safe Working Load and marked to indicate if it is used for lifting people
- Lifting operations are planned by a competent person, appropriately supervised and carried out safely
- Lifting equipment is subject to thorough examinations and reports by a competent person at specified times and inspected between thorough examinations, if appropriate.

2.5.10   The Construction, Design and Management Regulations

Many works which take place in the outdoor lighting environment either form part of a construction site or are, in themselves, a construction site. Construction work has a particularly poor record in terms of accidents and ill health, and investigations throughout Europe into the fundamental causes of accidents established that poor design and planning accounted for some 63 per cent (poor design 35 per cent, poor planning 28 per cent), whilst poor construction practice accounted for the remaining 37 per cent.

As a result of these figures and the continuing poor record of construction works, a number of European Directives were formulated and these have produced the Construction, Design and Management (CDM) Regulations and also other relevant Regulations such as the Construction (Health, Safety and Welfare) Regulations.

The aim of these Regulations is to improve the overall management and coordination of health, safety and welfare throughout all stages of a construction project and to place duties on all those who can contribute to health and safety.

The CDM Regulations apply in their entirety where the work is notifiable, that is it is construction work which will last for more than 30 days or 500 person days or there will be 5 or more persons on site at any one time or demolition or dismantling is involved. Whilst it is not intended that the CDM Regulations apply to term maintenance contracts (except where individual works therein fall within the scope of the CDM Regulations), some organisations have adopted the CDM Regulations as a matter of policy, applying the principles to all works. It should be noted in particular that Regulation 13 which sets out the duties of designers applies even where the rest of the CDM Regulations do not apply.

The CDM Regulations define five key parties and their duties and introduces the concepts of the Health and Safety Plan and the Health and Safety File. The five key parties are

1 The Client
2 The Designer(s)
3 The Planning Supervisor
4 The Principal Contractor
5 The (Other) Contractor(s).

The Client's role is essentially to appoint a competent Planning Supervisor and Principal Contractor and to provide Health and Safety information to the Planning Supervisor; the Designer(s) have to give due regard to Health and Safety in their design work and provide information; the Planning Supervisor notifies HSE of the project, ensures Designers co-operate with each other, advises the client (e.g. on competency) and ensures both the pre-tender Health and Safety Plan and the Health and Safety File are prepared.

It is the Principal Contractor's duty to develop the construction phase of the Health and Safety Plan before work starts, to ensure co-operation between Contractors and to provide information to the Planning Supervisor as well as ensuring that only authorised persons are permitted on site; the (Other) Contractor(s) should co-operate with and provide Health and Safety information to the Principal Contractor.

The Health and Safety Plan consists of two phases. The first is the pre-tender Plan which contains information mainly from the Client and Designer and which should focus on Health and Safety, provide sufficient information at the tender stage so as not to disadvantage any particular bidder and also allow the proper assessment of tender submissions. This is then developed into the Construction Phase of the Health and Safety Plan which contains the organisation, arrangements and standards for ensuring that risks are minimised during the works.

The Health and Safety File is, in essence, a Construction and Maintenance File or an Operating and Maintenance Manual which has been extended to alert persons responsible for the structure in the future to risks that must be managed during maintenance, repair, renovation and demolition. It is the Client's responsibility to ensure that the Health and Safety File is available for use after the project is complete.

2.5.11   The Construction (Health, Safety and Welfare) Regulations

The Construction (Health, Safety and Welfare) Regulations (CHSW Regulations) explain the detailed ways of working in construction activities and aim to protect the health, safety and welfare of everyone who carries out construction work as well as other people who might be affected by the work.

The key Regulations place duties on employers and those in control of construction work to

- Ensure a safe place of work and safe means of access to and from that place of work (Regulation 5)
- Prevent falls from height (particularly over 2 m) by physical precautions, or where this is not possible, provide equipment that will arrest falls (e.g. fall arrest equipment should always be worn when in the bucket of a Mobile Elevated Work Platform); erect access equipment under the supervision of a competent person; and ensure there are criteria for the proper use of ladders (Regulations 6, 7)
- Take steps to prevent people being harmed by falling materials or objects (Regulation 8)
- Prevent accidental collapse of new or existing structures or those under construction; ensure the dismantling or demolition of any structure is planned and supervised by a competent person (Regulations 9, 10, 11)
- Prevent people or vehicles from falling into excavations; prevent collapse of ground both in and above excavations; identify risks from underground cables and other services (Regulations 12, 13)
- Take steps to prevent people from falling into water or other liquid; ensure that personal protective equipment and rescue equipment is available in the event of a fall (Regulation 14)
- Ensure construction sites are organised so that both pedestrians and vehicles can move safely; prevent or control the unintended movement of any vehicle (Regulation 15)
- Ensure construction activities, where training, technical knowledge or experience is necessary to reduce risks, are carried out only by people who meet these requirements or who are supervised by people who meet the requirements (Regulation 28).

2.5.12   The Electricity at Work Regulations

Usually, outdoor lighting will involve the use of electricity. Electricity whilst being a convenient form of energy can and does kill, and even non-fatal electric shocks can cause serious damage – for example when falling from a height or working within potentially explosive atmospheres. The Electricity at Work (EAW) Regulations together with additional guidance such as HSE Guidance Notes, BS 7671 – Requirements for Electrical Installations (formerly known as the IEE Wiring Regulations) and the ILE Code of Practice for Electrical Safety in Highway Electrical Operations attempt to create a framework to avoid hazards and to minimise risks arising from the use of electricity and electrical equipment.

The documents referred to above are quite specific and it is important that both the scope of the documents and the definitions of the terms used within them are understood. The EAW Regulations cover every type of electrical equipment from, for example, a battery-powered hand lamp to a 400-kV overhead line and aim to minimise the danger resulting from electric shock, burns, explosion or arcing. The Regulations apply to 'duty holders' which includes employers and the self-employed and also recognises the level of responsibility which many employees within the electrical profession take on inasmuch as the amount of control that they might exercise over the work being carried out.

The key Regulations place requirements on the duty holders to ensure that

- All systems are of such construction as to prevent Danger and are maintained so as to prevent Danger (Regulation 4)
- Every work activity is carried out so as to prevent Danger; any equipment provided for protecting persons is suitable, maintained and properly used (Regulation 4)
- Electrical Equipment shall be of such construction or as necessary protected as to prevent Danger from exposure to mechanical damage; effects of weather, natural hazards, temperature or pressure; any flammable or explosive substance (Regulation 6)
- All Conductors in a System shall be suitably covered with insulating material and as necessary protected or have such precautions taken (e.g. by being placed out of reach) so as to avoid Danger (Regulation 7)
- Precautions shall be taken (e.g. by earthing or other means) to prevent Danger arising when any Conductor (other than a Circuit Conductor) becomes charged (Regulation 8)
- Suitable means (including where appropriate methods of identifying circuits) shall be available for cutting off the supply of electrical energy and for the isolation of any electrical equipment; adequate precautions shall be taken to prevent Electrical Equipment which has been made dead from becoming electrically charged (Regulations 12, 13)
- No person shall be engaged in any work activity on or so near any live conductor (other than one covered with insulating material) that Danger may arise unless it is unreasonable for it to be dead and it is reasonable for him to be at work on or near it and suitable precautions are taken to prevent injury (Regulation 14)
- Adequate working space, means of access and lighting shall be provided (Regulation 15)

- No person shall be engaged in any work activity where technical knowledge or experience is necessary to prevent Danger or Injury unless he possesses such knowledge or experience or is under such degree of supervision as may be appropriate (Regulation 16).

The EAW Regulations also provide a defence in any criminal proceedings for an offence consisting of a contravention of its absolute requirements under Regulation 29 in that the accused must show that he took all reasonable steps and exercised all due diligence.

2.5.13 Electricity Safety, Quality and Continuity (ESQC) Regulations

These Regulations, produced by the Department of Trade and Industry, replace the Electricity Supply Regulations. They were produced specifically to provide a framework which would allow competition in connections and as such introduce wider definitions and have wider applicability than the Electricity Supply Regulations. In particular the term 'Distributor' now includes organisations, such as local authorities, who distribute electricity through their own overhead or underground cables.

Whilst many of the health and safety requirements are mirrored in general terms in other legislation, the ESQC Regulations place specific duties on Distributors to

- Keep and maintain maps or plans showing underground cables
- Place suitable and sufficient warning notices to indicate presence of overhead cables
- Prevent so far as is reasonably practicable unauthorised access to overhead cables.

2.5.14 Lighting and Health and Safety Legislation

The foregoing sets out the general health and safety legislation which must be considered when dealing with outdoor lighting. It is worth repeating the fact that the general requirements of the HASAWA and the MHSWR require employers and self-employed persons to assess the risks from lighting and to put measures in place to ensure that their lighting is safe and does not endanger the health and safety of people who might be affected by them. In addition, there is a requirement on lighting manufacturers and suppliers to ensure the products supplied by them are safe, and to provide instructions on using and maintaining lighting.

Specific requirements for lighting are laid out in the Workplace (Health, Safety and Welfare) Regulations for both lighting and emergency lighting; in the PUWER – which requires that any place where a person uses work equipment shall be suitably and sufficiently lit and places a requirement to ensure that work equipment such as lighting complies with the relevant UK implementation of EC Directives; in the EAW Regulations – which has a number of requirements listed earlier; in the CHSW Regulations – which covers issues on construction sites such as emergency lighting, adequate natural or artificial lighting, and the colour of any artificial lighting used for traffic routes; and in the Control of Substances Hazardous to Health Regulations – which requires employers and the self-employed to protect people against health risks arising from hazardous substances to which they might be exposed during the installation, maintenance and disposal of lighting components and systems. In addition to this legislation, there are specific guidance documents produced by the HSE and others, which are applicable.

2.5.15 BS 7671 – Requirements for Electrical Installations (The IEE Wiring Regulations)

BS 7671 is sometimes referred to as 'deemed to comply' regulations in that the HSE regards compliance with this standard as likely to achieve compliance with the relevant aspects of the EAW Regulations. BS 7671 contains detailed regulations referring in particular to protection for safety, the selection and erection of equipment and both initial and periodic inspection and testing. In addition there is a section covering special locations, which includes Highway Power Supplies and Street Furniture which modifies or adds to the general requirements of BS 7671.

The IEE has also published a series of Guidance Notes (see Bibliography) which seek to provide further guidance and clarification in respect of the application of BS 7671 and it is within these that detail figures in respect of, for example, the frequency of periodic testing are contained.

2.5.16 The ILE Code of Practice for Electrical Safety in Highway Operations

The ILE Code of Practice together with the other ILE Technical Reports and the NVQ Occupational Standards in Public Lighting Installation and Maintenance provide a framework specific to the Outdoor Lighting environment which enables engineering judgement to be exercised against a background of competency and detailed guidance.

The ILE Code of Practice covers safe working measures, training, and the installation of cables, lighting columns and other street furniture. In addition both emergency and temporary situations are covered, with the latter being included in part within the IEE Guidance Notes on Special Locations. Sample Test and Inspection Certificates are also included as a template from which organisation-specific forms can be developed as required.

The ILE publications together with the ILE-led NVQ Occupational Standards complete the guidance that is needed to enable those involved within the field of Outdoor Lighting to satisfy their obligations under the many items of Health and Safety legislation which apply in this area.

2.5.17 National (Scottish) Vocational Qualification NVQ/SVQ

This nationally recognised qualification allows for the assessment of competence to be undertaken generally in the workplace.

The qualification has been developed for three levels of NVQ/SVQ:

1 Level 2 – Operative
2 Level 3 – Electrician
3 Level 4 – Managerial/Supervisory.

The occupational standard developed by a group representing the industry is intentionally biased to Health and Safety, having mandatory 'units' to assess the competence of workers on safety matters in the workplace.

2.5.18 Waste management

Information on a selection of key items of legislation for waste management is given in Section 2.6.2.

2.6 Waste management

2.6.1 Introduction

2.6.1.1 General

The management of waste is not only the moral duty of the lighting professional but a statutory requirement. Processes should be environmentally friendly, safe, effective and carried out in as efficient a manner as possible. Old practices of disposal at licensed landfill sites are no longer morally acceptable or legal.

Lighting and electrical equipment such as lamps and control gear can contain toxic elements which must be treated and ideally recycled, if at all possible.

It is estimated that, in the UK alone, 12 000 tonnes of glass, 670 tonnes of metal, 350 tonnes of plastic, 100 tonnes of phosphor and 0.5 tonnes of mercury are recoverable from light sources.[12]

The education of the next generation of lighting professionals about the legal requirements and environmental advantages of considering 'green' issues is the duty of all in the industry.

The reduction of toxic elements in the manufacture of equipment must be examined and where such elements are essential they should be recovered and reused. Energy used to facilitate such processes should also be recovered. Manufacturers should adopt a 'green philosophy' for the complete life of a product and its constituent parts and ideally this should include the real cost of recycling such products. The commercial difficulties of this philosophy should be recognised by clients, designers and installers.

Many manufactures still produce products without any real regard to the disposal consequences or sustainable resource use. The cost of disposal is then borne by the end-user and the consequences of pollution and depleting finite resource rest with society in general.

There have been many attempts to legislate to make producers responsible for waste but this has met with tremendous resistance from the business sector. The reluctance to recognise the true cost of safe disposal of manufacturer's products and absorb them into the cost of production is almost universal. The fear of losing competitive advantage is cited as the main reason. If sustainability of resources is to be achieved, this is a basic link that should be established.

In a global context the developed world consumes 80 per cent of resources used, for the benefit of only 20 per cent of the world's population.[13] This is largely done without producers accepting the consequences of extraction, processing and manufacturing of their equipment. Manufacturers, the waste management industry and all other stakeholders must rise to the challenge of sustainability and take a lead in meeting the many challenges to ensure improvements for future generations and the environment.

The 'green movement' has aided the understanding of waste as a sustainable resource and the introduction of European and UK legislation is now having a substantial impact on waste management practices.

The management of waste is one of the key themes of 'sustainable development'.[14] This concept originated from the 1992 United Nations Rio Conference on Environment and Development, and is defined as 'development that meets the needs of the present without compromising the ability of future generations to meet their own needs'.[15]

The four main aims of sustainable development as described in 'A Better Quality of Life',[16] the Government's Sustainable Development Strategy, are

1   Social progress which recognises everyone's needs.
2   Effective protection of the environment.
3   Prudent use of natural resources.
4   Maintenance of high and stable levels of economic growth and employment.

2.6.1.2 Waste hierarchy

The waste hierarchy, also known as the three Rs principle of waste management, are

1 Reduction
2 Re-use
3 Recovery.

These principles are becoming widely adopted with recognition from the UK Government coming in 1995 with the publication of 'Making Waste Work'.[17]

The basic principles should be utilised and implemented in the above order. The first principle of reduction should be implemented by manufactures as should re-use and recovery which are also the responsibility of the end-user. Energy is also included in this philosophy and the use of landfill sites and incineration without energy recovery should only be used as a last resort. All waste recycling and disposal solutions should be considered.

2.6.1.3 Best practical environmental option (BPEO)

The BPEO is used by planning authorities and industry as guidance when considering proposed waste management facilities. Proximity principles can be used in conjunction with the waste hierarchy to achieve the BPEO.

The Royal Commission on Environmental Pollution (RCEP) defined the BPEO as 'the outcome of systematic and consultative decision making procedure which emphasises the protection and conservation of the environment across land, air and water'. The BPEO procedure establishes, for a given set of objectives, the option that provides the most benefits or the least damage to the environment as a whole, at acceptable cost, in the long term as well as in the short term.

2.6.2 Legislation

2.6.2.1 General

The following are notes on the UK legislation that is particularly relevant to waste management. For wider issues in legislation see Section 2.5.

2.6.2.2 Management of Health and Safety at Work Regulations

The Health and Safety Executive, under the MHSWR, prescribes safety standards for exposure limits to hazardous substances in air, water, etc., in the working and living environment.

As all discharge lamps contain toxic elements such as mercury, sodium and various other heavy metals, it is therefore essential that they be treated, before being disposed of, to remove or neutralise any of these elements.

2.6.2.3 The Control of Substances Hazardous to Health (COSHH) Regulations

The COSHH Regulations place duties on employers to protect employees and other persons who may be exposed to substances hazardous to health. These substances could be solids, liquids or gases which may be irritant, toxic, harmful and/or corrosive. The COSHH Regulations require that anyone employed on the disposal or treatment of discharge lamps, for example, must be fully trained in the methods of work and equipped with the necessary safety equipment to ensure their safety and the safety of anyone nearby.

2.6.2.4 Controlled Waste Regulations

Controlled Waste (Registration of Carriers and Seizure of Vehicles) Regulations set standards for the transportation of waste and require that companies carrying waste be registered under same. Companies collecting and transporting discharge lamps for recycling or the residue of crushed lamps are also required to comply with these regulations.

Exemptions to this regulation are companies carrying their own waste, unless it is construction and demolition waste, British Rail and charities.

2.6.2.5 Waste Electrical and Electronic Equipment (WEEE) Directive

The legal framework controlling the disposal of discharge lamps and tubes changed significantly with the introduction of the Waste Electrical and Electronic Equipment Directive in 2003. The directive makes recycling of lamps and tubes a mandatory requirement and the new legislation is in line with the UK Government objectives with regard to reducing landfill and environmental pollution.

The WEEE Directive strives to ensure that dangerous substances such as mercury are disposed of in a safe and responsible manner and to encourage the recycling of as many components as possible. In order to comply with the WEEE directive and ensure the safe, efficient and effective recycling of products, a specialist contractor may be considered to be the best option.

2.6.2.6 Special Waste Regulations

An imminent review of the Special Waste Regulations will, like the WEEE, change the requirements for the disposal of lamps. The review, by the UK Government, is likely to bring lamps and tubes under its scope in the near future and will be geared to reducing environmental pollution.

2.6.2.7 Waste Management Licensing Regulations

The storage of waste is covered by the Waste Management Licensing Regulations. Large quantities of discharge lamps can be accumulated at depots and the like, waiting processing or collection. The requirements as detailed in this regulation permit the storage of non-liquid waste at any place other than the premises where it is produced if the following conditions are satisfied:

- Waste is stored in a secure building or container(s).
- It does not at any time exceed $50\,m^3$ in total and is not kept for a period longer than 3 months.
- The person storing the waste is the owner of the building or container(s) or has the consent of the owner.
- The place where it is stored is not a site designed or adapted for the reception of waste with a view to it being disposed of or recovered elsewhere.
- Such storage is incidental to the collection or transport of the waste.

2.6.2.8 Environmental Protection Act

This Act requires, amongst other things, for waste holders to exercise a duty of care when disposing of material containing polychlorinated biphenyls (PCBs). This duty applies to any person who imports, produces, carries, keeps, treats or disposes of controlled waste or, as a broker, has control of such waste. It requires such persons to ensure that there is no unauthorised or harmful deposit, treatment or disposal of the waste, to prevent the escape of the waste from their control or that of any other person, to ensure that the transfer of waste is only to an authorised person and that a written description of the waste is also transferred.

In April 2000 Part 2(a) of the act came into force introducing a new regime for the regulation of contaminated land. Its main purpose is to provide an improved system for the identification of land that is posing unacceptable risks to health or the environment and for determining a remedy where such risks cannot be controlled by normal means.

The Environmental Protection Act uses the term 'best available techniques', where techniques include technological hardware and operational procedures. In general terms a best available technique for one process is likely to be suitable for a comparable process.

2.6.2.9 The Environmental Protection (Duty of Care) Regulations

These Regulations came into force on 1 April 1992 and require that the transferor and the transferee of waste complete and sign a transfer note at the same time as the written description of the waste is transferred. The transfer note must identify the waste in question and state its quantity, how it is stored, the time and place of transfer, the name and address of the transferor and the transferee, whether the transferor is the producer or importer of the waste and further additional information.

Another requirement is the necessity for the transferor and the transferee to keep written descriptions of waste and the transfer notes, or copies of them, for 2 years from the date of transfer. This documentation must be supplied to a waste regulation authority if requested by the authority.

2.6.2.10 European Landfill Directive

The European Landfill Directive is influential legislation which will effect how the waste industry operates for many years to come. The most significant requirement of the Directive is that each member state should draw up a strategy for a three-stage reduction in the quantity of biodegradable municipal solid waste sent to landfill. Quantities must be reduced as

- 75 per cent of the 1995 figure by 2006
- 50 per cent by 2009
- 35 per cent by 2016.

A 4-year extension has been given to all member states that send more than 80 per cent of their municipal waste to landfill.

2.6.2.11 Environmental Protection Regulations

The Environmental Protection (Disposal of polychlorinated biphenyls and other Dangerous Substances) (England and Wales) (Amendment) Regulations 2000 came into force on 1 January 2001 and imposed a prohibition on the holding PCBs and equipment containing them after 31 December 2000. Extensions may have been allowed if major replacement programmes could be demonstrated.

2.6.3 Responsible bodies

There are various bodies responsible for environmental management and hence waste management (for details of these bodies, see Legistation in Bibliography).

2.6.4 Waste

2.6.4.1 General

The appropriate management of waste is determined not only by the relevant legislation but also by the equipment concerned, the constituent parts of same and the chemicals and elements contained therein.

2.6.4.2 Lamps

2.6.4.2.1 GENERAL

The most hazardous elements found in lamps are mercury and sodium used in discharge types. The pressures that exist in the discharge tubes present further dangers due to the possible ejection of contaminants if broken. However, the small quantities of these toxic elements in each lamp means that they are not classified as special waste under the Special Waste Regulations. The waste products formed as a result of treatments such as crushing, etc., may, however, be classified as special waste and therefore need to be treated accordingly.

Small quantities of discharge lamps are often broken and treated manually. This can be a dangerous process due to the perils of flying glass, the possible inhalation of particles and powders and the risk of explosion of flammable gases. Safe and efficient disposal and treatment is best carried out in purpose-built mechanical crushers. Modern machines have the ability not only to crush lamps and neutralise or collect toxic elements but also to ensure that there is no escape of airborne contaminants to endanger operators, observers or the environment.

Specialist lamp recycling companies are now starting to appear which provide a comprehensive, fully automated service. Such systems recover toxic element and render glass, metal end caps and the like suitable for recycling.

All recycled materials should be recorded in documentation, which has a verifiable audit trail. This is particularly important to ensure that materials such as contaminated glass from discharge lamps are not reused in the food industry.

The Environment Agency produces Special Waste Explanatory Notes for Lamps containing mercury and sodium, Guidance Notes SWEN 047 and SWEN 047A respectively. The Lighting Industry Federation (LIF) produces a booklet, Technical Statement No. 10, which covers The Handling and Disposal of Lamps.

2.6.4.2.2 SUSTAINALITE SCHEME

The Lighting Industry Federation (LIF) in conjunction with the waste management industry has formed the SustainaLite scheme. This scheme has full government backing and will maintain a central register of accredited lamp recycling companies which have been thoroughly vetted and are continually monitored. The criteria of accreditation meet all the requirements of the WEEE Directive.

SustainaLite's primary objective is 'to promote best practice in the management and resource use of end-of-life gas discharge light sources and to establish and manage an accreditation scheme for those who manage end-of-life gas discharge light sources'. All companies accredited to the scheme will have to be accredited to ISO 9001 and/or ISO 14001.

In January 2002 the first companies to receive the SustainaLite Waste Management and Recycling Certificate of Accreditation were awarded their certificates. The Certificate of Accreditation recognises that these companies operate an ISO 9001 quality system incorporating the audit conditions drawn up by the British Standards Institution for the collection and recycling of gas discharge lamps. Also their certification is established by an independent body, which is registered for ISO 9001/2 by United Kingdom Accreditation Service (UKAS).

2.6.4.2.3 MERCURY (Hg)

Mercury is most commonly recognised as the shiny silvery metallic liquid found in thermometers. This heavy element is, however, found naturally in the environment in several forms.

Mercury can be combined with other elements to form inorganic compounds, can evaporate to form colourless, odourless mercury vapours and can combine with organic material to form organic compounds such as methylmercury (MeHg). The latter is produced primarily by bacteria and is the form which poses the greatest danger to the environment.

Exposure to high levels of elemental mercury vapour can result in nervous system damage including tremors, and mood and personality alterations. Exposure to relatively high levels of inorganic mercury salts can cause kidney damage. Adult exposure to relatively high levels of methylmercury through, for example, fish consumption can result in numbness or tingling in the extremities, sensory losses and loss of coordination. Exposure of the developing foetus through maternal intake of contaminated fish can result in neurological development abnormalities in cognitive and motor functions. Whether any of these symptoms actually occur, and the nature and severity of the symptoms, depend on the amount of exposure.

People are most likely to be exposed to metallic mercury due to mercury being released from dental fillings. However, the amount of mercury released from dental fillings is generally not considered to be high enough to cause adverse health effects.

In the case of the lighting professionals and operatives, exposure may also result from industrial processes or from breathing in air contaminated with vapours from metallic mercury spills. Mercury contained in the phosphor coatings of fluorescent and mercury lamps can be released during crushing. The removal of the dangers of inhalation of this and other chemicals can be achieved by the addition of water and the use of pressurised extraction systems. Such systems should remove the mercury for reuse and ensure that only well-filtered air and water is released into the environment.

Exposure and health risks may be determined by measuring the amounts of mercury in blood, urine, breast milk and hair. Over time, however, the body can rid itself of some contamination.

Liquid waste may be discharged into foul sewers as trade effluent, following treatment to reduce pH levels. The remaining waste such as glass and metal end caps can be disposed off at registered waste disposal sites where further recycling is likely to take place.

2.6.4.2.4 SODIUM (Na)

Drenching with copious amounts of water can treat sodium released from low- and high-pressure lamps during crushing. This does, however, lead to the release of hydrogen, which is a highly explosive gas. Care should therefore be taken to ensure that the process is only carried out in well-ventilated areas or outdoors. The other by-product of this chemical reaction is sodium hydroxide (NaOH). This is a corrosive solution, which needs to be neutralised by dilution and adjustment of pH to approved levels before it can be discharged as trade effluent to the foul water sewer. The pH and concentration will determine whether or not this substance should be treated as 'special waste'.

2.6.4.3 Capacitors, transformers, ballasts, circuit breakers and switch-gear

2.6.4.3.1 GENERAL

The safe disposal of capacitors, transformers, ballasts, circuit breakers and switch-gear should be considered as many of these components, especially if they have been in service for some time, may contain hazardous chemicals such as PCBs.

2.6.4.3.2 POLYCHLORINATED BIPHENYLS (PCBs)

PCBs are a family of organochlorine chemicals that are chemically stable, fire resistant and do not generate vapours easily. They are practically insoluble in water, but are soluble in oils and fatty substances. There are 209 congeners or forms of PCBs and commercially available PCBs are mixtures of congeners with varying chlorine contents.

PCBs were used extensively as insulators in electrical equipment such as capacitors, transformers, ballasts, circuit breakers and switch-gear.

It should be assumed that any capacitor or transformer manufactured before 1976 may contain PCBs unless there is information to the contrary. It is also possible that PCBs may be present in transformers and capacitors manufactured between 1976 and 1986.

PCBs are not considered to be acutely toxic to humans; however, repeated exposure can cause them to accumulate in the body. They can cause a skin condition called chloracne that produces pustules, blackheads and cysts. In animals they can also cause damage to the liver, and reduce the effectiveness of the immune system. Strong heating of PCBs in the presence of oxygen can lead to the formation of dibenzofurans, which may pose a greater hazard to human health – hence the need for safe, effective waste management.

PCBs can pass easily through the skin; therefore protective equipment should be worn if there is any possibility of skin contact. Any cuts or abrasions should be covered with dressings before putting on the protective clothing.

2.6.4.4   Lighting columns

All types of redundant lighting columns should be recycled where at all possible and previous practices of burying in landfill sites should be abandoned.

Concrete lighting columns should be crushed and the spoil recycled for use in the construction industry. Many companies now provide such a service.

Fibreglass lighting columns should be carefully dismantled to avoid the release of contaminants into the environment. Sections can be shredded and recycled to form fillers for plastic mouldings.

Chapter 3

Equipment

3.1 Introduction

In order to ensure a successful lighting project which is in harmony with the outdoor environment in which it sits, it is critical that appropriate equipment is selected. There are many influences to consider and reference should be made to Section 4.1.

The aim of this section of this chapter is to give generalised accounts of the principal types of equipment used in outdoor lighting and, as appropriate, their principles of operation, material characteristics and strengths and weaknesses.

As the effective maintenance of equipment is essential to prevent excessive deterioration, unwarranted energy consumption and increased maintenance costs, there is a section devoted to maintenance principles.

It should be recognised that the information provided can neither describe every available item of equipment nor be entirely current. However, judicious use of the principles contained herein should contribute to satisfactory outcomes.

3.2 Light sources

3.2.1 Introduction

The choice of light source for a particular application is fundamental in defining both the optical performance and more importantly the feel of the installation. Therefore considerable thought needs to be given to the decision.

In order that the correct choice of lamp is made, consideration should be given to all the relevant lamp characteristics and as such the choice is not necessarily that simple. The important characteristics are as given in the following sections.

3.2.2 Lamp characteristics

3.2.2.1 Luminous efficacy

The ratio of the luminous flux emitted by a lamp to the power consumed by the lamp and associated control gear. Effectively it is a measure of how efficient the conversion of electrical power into light output is, and is measured in lumens/watt (lm/W).

3.2.2.2 Correlated colour temperature

The colour of the light emitted by a 'near white' source is indicated by its CCT. This is a measure of the 'warmth' or 'coolness' of the light emitted by a source and is measured in Kelvin (K). The lower the kelvin value, the 'warmer' the colour of the light and vice versa. The effects of different colour temperature is illustrated in Figure 1.5.

3.2.2.3 Colour rendering

The ability of a light source to render colours of surfaces correctly is quantified by the CIE colour rendering group and the CIE general colour rendering index (R_a) as shown in Table 3.1. This index is based on how close a set of test colours are reproduced by the lamp under evaluation, relative to how they are reproduced by an appropriate standard light source of the same CCT. Perfect matching is given a value of 100.

3.2.2.4 Lumen output

Lumen output is the radiant power of visible light emitted from a source and is measured in lumens (lm).

Table 3.1 CIE colour rendering index groups

Colour rendering groups	CIE general colour rendering index	Typical applications
IA	$R_a \geq 90$	Critical colour matching
IB	$90 \geq R_a \geq 80$	Accurate colour judgements required for appearance
2	$80 \leq R_a \leq 60$	Moderate colour rendering required
3	$60 \leq R_a \leq 40$	True colour recognition of little significance
4	$40 \leq R_a \leq 20$	Not recommended for colour matching

3.2.2.5   Lamp lumen maintenance factor (lumen depreciation)

All light sources suffer lumen output depreciation with operating hours. The rate of depreciation is dependent on the lamp type, and is influenced by some of the following factors:

- Tungsten migration;
- Electrode deterioration;
- Emitter depletion;
- Mercury absorption into glass;
- Migration of light emitting material through the discharge tube.

The rate of depreciation is normally represented graphically and a typical example is shown in Figure 3.1.

3.2.2.6   Average life

Average life is the length of time at which 50 per cent of a large sample batch of lamps will have failed when tested under reference conditions, measured in hours. A typical survival curve is shown in Figure 3.2.

3.2.2.7   Burning position

Some lamps are only capable of being operated in certain positions. This may influence the lamp choice (e.g. some metal halide lamps can only be operated horizontally), and therefore the luminaire choice. Operating the lamp in the wrong burning position will almost certainly result in poor performance either in terms of life, colour or output.

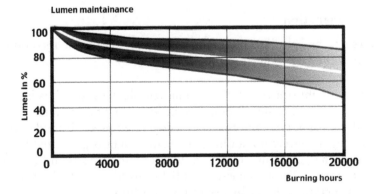

Figure 3.1 Typical lamp lumen maintenance curve.

Note: Curve is not representative of any specific lamp type, wattage or manufacturer.

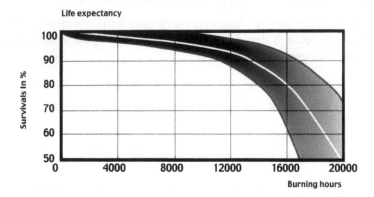

Figure 3.2 Typical lamp survival curve.

Note: Curve is not representative of any specific lamp type, wattage or manufacturer.

3.2.2.8   Strike and re-strike time

Tungsten and tungsten halogen lamps light up instantly and produce their full amount of light immediately. These lamps can be switched off and then immediately on again producing full light output. Fluorescent lamps light up within a few seconds after turning a switch on and under most circumstances produce the majority of their light output within 15 s. Almost all HID light sources take several minutes to reach full light output after ignition. Unless fitted with special control gear, once extinguished there will be a significant delay before these lamps will re-strike (between 30 s and 15 min).

3.2.3   Light generation

Light sources can be split into two 'families' depending on the method of light generation they use. They are as follows:

1   *Incandescent* light sources employ a tungsten filament, which is heated electrically and as a result incandesces (i.e. emits light at high temperatures). All tungsten GLS and tungsten halogen lamps are incandescent.
2   *Discharge* light sources do not have filaments. Light is generated by an electrical discharge passing through a tube containing a gas and various metallic elements. These elements emit light directly (e.g. high pressure sodium or metal halide) or by excitation of a fluorescent powder (e.g. fluorescent and induction lamps), or indeed a combination of these two processes (e.g. mercury comfort lamps).

A gaseous discharge is generally obtained by sending an electric current though a gas between two electrodes (or inducing the current in the case of induction lamps). In the case of an a.c. discharge, the electrodes alternately serve as anode or cathode. The actual carriers of the electric current in the gas between the electrodes are the electrically charged particles, positive ions and negative free-moving electrons.

Some of these electrons will collide with neutral atoms of the gas on their way from one electrode to the other. If the speed of these collisions is moderate, the collision will temporarily eject one of the electrons of the gas to a higher orbit (excitation). These collisions result in the emission of electromagnetic radiation, which takes place when the excited electron falls back to its previous orbit. The wavelength of the radiation is dependent on the composition of the gas and the pressure within the discharge tube.

3.2.4 Principal lamp type characteristics

3.2.4.1 Incandescent

Commonly known as GLS and TH; ILCOS codes – IA, IB, IN, IRR, IPAR, HS, HD, HR or HM.

The use of tungsten and tungsten halogen lamps in outdoor applications has declined considerably as more efficient light sources have become available. However, there are still a number of applications where tungsten lamps are the most suitable choice. These would include some decorative applications where continuous dimming is required (perhaps in combination with colour filters for use in a colour cross-fade system), or perhaps where small light sources that do not require either control gear or transformers, are used.

Tungsten halogen lamps, whilst having a slightly higher efficacy, are still limited in their use. Again the main applications would tend to be decorative floodlighting or accent lighting of smaller objects or monuments. The most frequently used types would be PAR reflectorised versions, or linear lamps. Low voltage (6 V, 12 V or 24 V) halogen lamps could be used in some applications, provided there is room to house the associated transformer, either within the luminaire or remotely. There are associated advantages with having a remote transformer, that is the wiring to the luminaire is at low voltage and is therefore intrinsically safer.

Both standard incandescent and halogen sources are sometimes used for PIR-controlled security lighting on account of their instantaneous lighting output. A summary of incandescent lamp characteristics are shown in Table 3.2.

Table 3.2 Summary of tungsten lamp characteristics

	Tungsten	Tungsten halogen
Commonly known as	GLS	TH
ILCOS code	IA, IB, IN, IRR, IPAR	HS, HD, HR or HM
Wattage range (W)	15–1000	5–2000
Luminous efficacy (lm/W)	7–14	12–22
Colour temperature (K)	2200–2800	2700–3200
CRI	100	100
Average rated life (hours)	1000	1000–5000
Burning position	Most – universal	Most – universal
	Linear – horizontal	Linear – horizontal
Run-up time	Instant	Instant
Re-strike time	Instant	Instant

3.2.4.2 Low pressure mercury

Commonly known as fluorescent lamps; ILCOS codes – FD and FS.

Electrodes at either end of the tube allow a current to be passed through the mercury vapour. This excites the mercury atoms, which emit UV radiation. The glass tube is internally coated with a phosphor powder, which converts this UV radiation into visible light. Triphosphor powders provide higher efficacy, better colour rendering performance and superior lumen maintenance.

There is a wide range of fluorescent lamps available that make them suitable for a wide range of applications. Whilst the largest application area is undoubtedly indoor retail, office and industrial lighting, many fluorescent lamps are used in outdoor applications. Linear fluorescent lamps are produced in lengths from 150 to 2400 mm, in a variety of tube diameters.

T12 (38 mm diameter) lamps, whilst they are unfortunately still being used in other applications, have no place in outdoor installations as they have a very poor efficacy and are unsuitable for operation on energy-efficient electronic ballasts.

The more efficient T8 (26 mm diameter) lamps are a better option where linear fluorescent lamps are required. These lamps are available in a range of colour temperatures and colour rendering indices. In general, triphosphor lamps should be used as they offer superior performance in efficacy, colour rendering and lumen maintenance.

T5 (16 mm diameter) miniature lamps, 4–13 W, are used in a wide range of sign lighting applications (i.e. small illuminated traffic signs). A new generation of TL5 (16 mm diameter) lamps, 549–1449 mm, and circular versions are now available and offer efficacies in excess of 100 lm/W. These lamps are optimised to operate in an ambient temperature of 35 °C; therefore care should be exercised when used externally (i.e. the luminaire should be designed so that the lamp is operated at the intended temperature).

The use of compact fluorescent lamps is increasing rapidly in the outdoor sector. Originally lamps were used in illuminated bollards, traffic signs and bulkhead-type luminaires. However, with the increase in popularity of white light, applications have widened considerably. There is now an impressive range of luminaires designed for compact fluorescent lamps of all types, suitable for both functional and decorative applications.

There are various formats of compact lamps including 2, 4 and 6 'limb' versions, with many being available in 2- and 4-Pin options. 2-Pin lamps have the starter in the base and therefore simply require the use of a ballast. 4-Pin lamps can be used with high frequency control gear, with associated benefits in terms of energy efficiency and voltage stability.

Most compact fluorescent lamps are only available in triphosphor colours. All fluorescent lamps lumen outputs are affected by ambient temperature. With standard 'cold spot' lamps, lumen output will decrease as ambient temperature moves away from the optimum (either above or below). However, 'amalgam' compact fluorescent lamps will provide their maximum light output over a wider range of temperatures and are therefore, in many situations, more suitable for use outdoors.

An amalgam-based fluorescent lamp ensures a near-constant mercury pressure over a wide temperature range, by 'releasing' and 'absorbing' the mercury at different temperatures. This near-constant mercury pressure equates to near-constant light output. A summary of fluorescent lamp characteristics are shown in Table 3.3.

3.2.4.3 Induction

Commonly known as QL, Endura or Genura.

Induction lamps are a modern advance in conventional fluorescent lamp technology. Rather than using an electrode at either end of the discharge tube, the ionisation of the gas is brought about by the induction of a high frequency electromagnetic field. Without electrodes to fail, the life of these lamps is very long (in excess of 60 000 h to 10 per cent failures). They are available in two basic shapes: globes and rectangular 'doughnut' shaped. Due to their impressive lifetime and lumen maintenance, induction lamps are highly suited to many outdoor applications. The obvious advantage is in extended lamp replacement periods and therefore lower maintenance costs. In addition, lamps can be incorporated into 'sealed for life' luminaires. Whilst each of the commercially available induction lamp systems have their own advantages and disadvantages (e.g. Endura is well suited to tunnel luminaires, QL well suited to luminaires with circular distributions and Genura well suited as a retrofit), the overall 'concept' of induction lamps is

Table 3.3 Summary of fluorescent lamp characteristics

	Fluorescent T12	Fluorescent T8	Fluorescent TL'5	Compact fluorescent
Commonly known as	T12	T8	TL5	CFL
ILCOS code	FD	FD	FD	FS
Wattage range (W)	20–125	10–70	14–80	5–55
Luminous efficacy (lm/W)	Halophosphate 50 Triphosphor 80	Halophosphate 50 Triphosphor 80 Multi-band 65–75	Triphosphor 95	Triphosphor 80 Multi-band 85
Colour temperature (K)	2700–6500	2700–6500	2700–6500	2700–6500
CRI	Halophosphate 50–75 Triphosphor 85	Halophosphate 50–75 Triphosphor 85 Multi-band 90+	Triphosphor 85	Triphosphor 85 Multi-band 90+
Average rated life (hours)	8000–10 000	Halophosphate 10 000–12 000 Triphosphor 16 000–20 000 Multi-band 10 000–12 000	20 000	9000–10 000
Burning position	Universal	Universal	Universal	Universal
Run-up time	Few seconds	Few seconds	Few seconds	Few seconds
Re-strike time	Few seconds	Few seconds	Few seconds	Few seconds

that they are rather costly, in practice this is no longer the case. A summary of induction lamp characteristics are shown in Table 3.4.

3.2.4.4 Low pressure sodium

Commonly known as SOX; ILCOS code – LS.

The low pressure sodium (SOX) type is a discharge lamp using only sodium vapour at a low pressure in a glass arc tube, housed in a tubular glass outer envelope. A SOX lamp comprises of a 'U' shaped discharge tube with supports, a gas filling, electrodes, outer envelope and a lamp cap.

The U-shaped discharge tube is constructed from ply-glass. This is soda-lime glass coated on the inside with a sodium-resistant borate glass. A great deal of care is taken when forming the discharge tube to prevent stress fractures and other weaknesses that could result in premature lamp failure. The construction of the lamp must also ensure that sodium is not allowed

Table 3.4 Summary of induction lamp characteristics

	Induction lamps
Commonly known as	QL, Endura, Genura
ILCOS code	
Wattage range (W)	55–165
Luminous efficacy (lm/W)	80
Colour temperature (K)	2700–6000
CRI	85
Average rated life (hours)	Up to 60 000 (10% failures)
Burning position	Universal
Run-up time	Instant
Re-strike time	Instant

to come into contact with the lead-in wires, or base of the electrode construction as this will cause premature failure.

Much of the SOX lamps design is focussed towards obtaining and maintaining high thermal efficiency and therefore high efficacy. Due to the fact the SOX lamps are physically quite large and the wall temperature must be maintained at 260 °C, it is essential that the outer bulb provide very good thermal insulation. As such the inner surface of the outer envelope is coated with a thin layer of heat-reflecting indium oxide (In_2O_3). This ensures that a high lamp efficacy is maintained. The outer envelope is evacuated.

Because sodium has a melting point of 98 °C, there is negligible sodium vapour in a cold SOX lamp. To avoid the need for excessively high ignition voltages, neon (and 1 per cent by weight of argon) is added to the discharge tube as a buffer gas and Penning mixture. In this way SOX lamps can be ignited on a voltage peak of between 500 and 1500 V. SOX and the more energy-efficient SOX-E versions were very widely used as a functional street lighting lamp and for security lighting. This popularity was largely due to the impressive efficacy of the source. However, the disadvantage with SOX lamps is their poor colour rendering index. The characteristics of low pressure sodium lamps are tabulated in Table 3.5.

3.2.4.5 High pressure sodium

Commonly known as SON; ILCOS codes – ST or SE.

The high pressure sodium (SON) lamp is a discharge lamp using sodium–mercury amalgam at a high pressure in a ceramic arc tube housed in a tubular or elliptical glass outer envelope. Due to the high temperatures and pressures found inside the SON lamp, its construction is quite different from the SOX lamp. A SON lamp comprises of a discharge tube with

Table 3.5 Summary of low pressure sodium lamp characteristics

	Low pressure sodium
Commonly known as	SOX, SOX-E, SOX Plus
ILCOS code	LS
Wattage range (W)	18–180
Luminous efficacy (lm/W)	SOX/SOX Plus 160
	SOX-E 190
Colour temperature (K)	1750
CRI	N/A
Average rated life (hours)	16 000
Burning position	18 W SOX-E – vertical base-up
	SOX – horizontal
Run-up time	12–14 min
Re-strike time	Instant to 15 min

supports, a gas filling, electrodes and feedthroughs, outer envelope, thermal switch and/or starting aid (where fitted) and a lamp cap.

The discharge tube of a SON lamp is made from sintered polycrystalline alumina (PCA). PCA is translucent, gas-tight and impervious to hot sodium vapour. The discharge tube is constructed from a hollow cylinder, which is closed at the ends with either niobium or alumina.

Xenon is added to the gas filling as a starting aid. A disadvantage of xenon is that it necessitates the use of a high voltage peak to start the xenon discharge (there being no known suitable Penning mixture).

Increasing the xenon pressure within the discharge tube results in a higher luminous efficacy. However, this also raises the ignition voltage required, beyond that which a standard ignitor can provide. Higher xenon pressures are employed in SON Plus lamps (increased lm/W). Reliable starting is achieved with the use of an additional starting aid, either a bi-metallic strip with an external antenna or the more reliable integrated antenna.

Internal ignitor lamps (i.e. not requiring an external ignitor) have a glow-discharge starter (similar in principle to a fluorescent starter switch) located near the base of the lamp. These lamps still require a ballast to limit the current supplied to the lamp.

In the colour-improved SON Deluxe or Comfort lamps the working pressure of the sodium is increased approximately four times. In the White SON lamp the sodium pressure has been further increased and a slightly different construction is employed.

Tubular lamps tend to have clear outer envelopes and are more suited to precise optical systems. Elliptical lamps can be clear or can have a diffusing coating added to reduce their brightness. Coated lamps are suited to more general optical systems. Regardless of shape, the outer envelopes are usually evacuated.

There are a number of types for use in outdoor applications, including functional and decorative street lighting, security lighting and floodlighting:

- SON-E is the standard elliptical lamp
- SON-T is the clear tubular version offering better optical control
- SON-DL (or comfort) offers improved colour rendering
- SON-R is a reflector version
- SON-TD is a tubular, double-ended lamp
- SON-Plus offers an improvement in efficacy, therefore a higher light output for the same input power.

In general, where standard SON light (i.e. R_a 20) is required, the SON Plus lamp should be used due to its superior efficacy, and therefore lower environmental impact. Some manufacturers are now producing mercury free lamps that will operate on standard control gear circuits. These lamps have approximately the same efficacy as standard SON lamps. The characteristics of the high pressure sodium lamp are tabulated in Table 3.6.

3.2.4.6 White high pressure sodium

Commonly known as white SON; ILCOS code – STH.

A less widely used source in the outdoor environment is the White SON lamp. This is a relatively compact lamp offering a warm colour temperature and excellent colour rendition, particularly of warm colours. Applications are generally limited to decorative floodlighting and amenity lighting. Characteristics of white high pressure sodium lamps are tabulated in Table 3.7.

A compact version of the standard White SON lamp has recently been introduced – miniWhite SON. This lamp offers the same colour quality in the dimensions of a 150 W CDM-T lamp.

Table 3.6 Summary of high pressure sodium lamp characteristics

	High pressure sodium		
Commonly known as	SON	SON DL/Comfort	SON Plus
ILCOS code	ST, SE	ST, SE	ST, SE
Wattage range (W)	50–1000	150–400	70–600
Luminous efficacy (lm/W)	100	85	120
Colour temperature (K)	1950	2150	1950
CRI	20	60	20
Average rated life (hours)	20 000	20 000	20 000
Burning position	Universal	Universal	Universal
Run-up time	6–12 min	6–12 min	6–12 min
Re-strike time	30 s–5 min	30 s–5 min	30 s–5 min

Table 3.7 Summary of white SON lamp
characteristics

	White SON
Commonly known as	SDW-T/TG
ILCOS code	STH
Wattage range (W)	35–150
Luminous efficacy (lm/W)	35
Colour temperature (K)	2500
CRI	83
Average rated life (hours)	10 000
Burning position	Universal
Run-up time	2 min
Re-strike time	1 min

3.2.4.7 High pressure mercury

Commonly known as MBF or HPLN; ILCOS codes – QE, QR or QB.

A discharge lamp using only mercury vapour at a high pressure in a quartz arc tube housed in an elliptical glass outer envelope.

A high pressure mercury lamp comprises of a discharge tube with supports, a gas filling, electrodes, outer envelope, bulb coating and a lamp cap.

The discharge tube is made from quartz and is pinched-sealed at each end. Running through these seals are molybdenum feedthroughs onto which main and auxiliary electrodes are attached. The discharge tube is filled with an inert gas (to aid starting and increase electrode life) and an accurate dose of distilled mercury.

High pressure mercury lamps do not require an external high voltage ignitor. This is primarily due to the use of an auxiliary electrode, which is simply a piece of molybdenum or tungsten wire positioned close to one of the main electrodes and connected to the other through a 25-kΩ resistor.

The outer envelope will usually contain an inert gas at atmospheric pressure, shielding the discharge tube from changes in ambient temperature. The inner surface of the outer envelope of standard mercury lamps is coated with a white phosphor to improve the colour rendering and the light output. Improved colour rendering mercury lamps (i.e. deluxe or comfort versions) use a superior phosphor coating that gives the lamp a lower colour temperature, and a higher colour rendering index and efficacy.

High pressure mercury lamps are generally regarded as 'old fashioned' although many existing installation use these lamps. Standard lamps are predominantly used in functional street lighting luminaires. Colour improved Deluxe (or Comfort) versions tend to be used in preference to standard mercury lamps due to their superior colour properties and improved

Table 3.8 Summary of high pressure mercury lamp characteristics

	High pressure mercury	Mercury comfort	Mercury blended
Commonly known as	MBF, HPL-N, HPL-R	MBF-SD, HPL-Com	ML-L, ML-R
ILCOS code	QE, QR	QE	QB
Wattage range (W)	50–700	80–400	100–500
Luminous efficacy (lm/W)	50	55	10–26
Colour temperature (K)	4100–4300	3500	3600
CRI	40–50	50–55	50
Average rated life (hours)	20 000	20 000	6000
Burning position	Universal	Universal	Universal
Run-up time	5 min	5 min	5 min
Re-strike time	10 min	10 min	5 min

efficacy. Reflectorised lamps are also available as the so-called 'mercury-blended lamps'.

Mercury-blended lamps incorporate a tungsten filament mounted around the discharge tube. This means the lamp can be used without a ballast as the filament acts as the current limiting device. This has the effect of reducing the efficacy, shortening the life, but increasing the colour rendering index.

The light from all of these sources can have a green 'tinge', giving the environment a slightly odd appearance. This effect coupled with a relatively poor efficacy has tended to result in installations being replaced with other sources. The characteristics of high pressure mercury lamps are tabulated in Table 3.8.

3.2.4.8 Metal halide

Commonly known as MBI or HPI; ILCOS codes – MT or ME.

The metal halide type is a discharge lamp that uses mercury vapour together with metal halide additives at a high pressure in a quartz or ceramic arc tube (see Section 3.2.4.9), contained within a glass or quartz glass outer envelope.

Most metal halide lamps are broadly similar in construction; however, due to differing technologies and construction principles it is not possible to generalise on construction techniques. There is a wide variety in the shape, size and precise material used in the construction of the discharge tube. Due to their complex nature and the action of the metal halides within the discharge tube many different electrode constructions exist and different shaped outer envelopes and coatings are also used.

Though the discharge tube contains mercury, it does not act as a light generator, instead the metal halides perform this role. Each combination of metals used generates a light of specific characteristics.

Metal halide lamps offer very good colour rendering and are widely used in sports lighting and exterior floodlighting. Now the performance of the lower wattage lamps is improving, they are finding their way into functional and decorative street lighting and amenity lighting applications. They are also widely used for area floodlighting.

There is an enormous range of lamp versions available and care should be taken when matching a lamp to control gear, or selecting a lamp for use in an existing installation, as compatibility could be an issue. Some lamps operate on mercury gear (low lamp current and low voltage ignition ($\sim$750 V)) and some operate on SON gear (higher lamp currents and high ignition voltages ($\sim$4.5 k V)).

Another area for caution is that of burning position. Most standard metal halide lamps are optimised for either horizontal or vertical operation. Again care should be taken when selecting a specific lamp and it is always advisable to consult the relevant manufacturer. The characteristics of metal halide lamps are tabulated in Table 3.9.

3.2.4.9   Ceramic metal halide

Commonly known as CDM, HCI or CMH; ILCOS code – MN.

Ceramic metal halide lamps offer advantages in terms of efficacy and colour stability over their quartz counterparts. The range is limited to low wattage versions (35–150 W), however, it is expected that the range will expand into medium wattages (i.e. 250 and 400 W) in the near future.

There are a number of problems associated with quartz metal halide lamps that are insurmountable using quartz technology. Discharge tubes

Table 3.9 Summary of metal halide lamp characteristics

	Metal halide
Commonly known as	MBI, HPI
ILCOS code	MT, ME
Wattage range (W)	50–3500
Luminous efficacy (lm/W)	80–100
Colour temperature (K)	2700–6000
CRI	65–90
Average rated life (hours)	4000–20 000
Burning position	Various
Run-up time	Various
Re-strike time	Various

constructed from a ceramic material known as PCA allow many of these problems to be overcome.

In many other ways the principle of operation is similar to quartz-based lamps, that is a discharge tube filled with a small amount of mercury and specific metal halides. Different manufacturers use slightly different techniques to construct the discharge tube; however, the critical factor is to ensure a gas-tight seal (even at high temperatures and pressures) between all the parts of the discharge tube. The main advantages of ceramic discharge metal halide lamps are improved colour stability, increased life and higher efficacy.

In addition to the obvious applications of amenity and decorative floodlighting, numerous functional street lighting installations have also been completed. It is anticipated that their use in street lighting will expand as manufacturers address the issues of life and lumen maintenance.

The range of lamp formats is wide and expanding; however, the most popular types for exterior applications have proved to be

• Single-ended tubular G12 cap (for functional street lighting and performance floodlights)
• Single-ended tubular E27/E40 cap (for functional street lighting)
• PAR versions (for compact floodlights)
• Double-ended RX7s cap versions (for wash floodlights).

The characteristics of ceramic metal halide lamps are tabulated in Table 3.10.

3.2.5 Other light sources

3.2.5.1 Light emitting diodes (LEDs)

When current flows across the junctions of certain solid-state semiconductor devices, light is emitted.

Table 3.10 Summary of ceramic metal halide lamp characteristics

	Ceramic metal halide
Commonly known as	CDM, HCI, CMH
ILCOS code	MN
Wattage range (W)	35–150
Luminous efficacy (lm/W)	90
Colour temperature (K)	3000–4200
CRI	85–95
Average rated life (hours)	10 000–15 000
Burning position	Most – Universal Double ended – Horizontal
Run-up time	3 min
Re-strike time	15 min

The production involves a process known as *epitaxy* in which crystalline layers of different semiconductor materials are grown on top of one another by chemical vapour deposition. Electrons and holes (electron vacancies) flow in opposite directions across the p–n junction so formed, until equilibrium is reached. When a potential (voltage) is applied electrons flow across the junction. As the electrons and holes (electron vacancies) recombine, either light (photons) or heat is emitted, dependent on the energy transition in electron volts. The wavelength of the light is determined by the relationship

$$\lambda = \frac{1240}{\Delta E} \, \text{nm}$$

where ΔE is the energy transition.

The composition of the materials in the semiconductor determines the energy transition. Typical configurations for light emission are given in Table 3.11.

In addition to wavelength, internal photon absorption rate and surface reflection (back into the material) also influence the efficiency of the materials. Surfaces are shaped to reduce internal reflection. Development of surface phosphor application (like a fluorescent lamp) is expected to increase efficiency and widen the range of useful materials. In some applications phosphors are used to produce 'white' light.

Light emitting diodes are small, lightweight, durable, have long life and produce light almost instantaneously. They are not affected by frequent switching, and can easily be dimmed. They operate at low voltage, which is an advantage in some situations, but a disadvantage in others where a transformer would be required.

The light generating chip is typically $0.25 \, \text{mm}^2$. The plastic encapsulant and lead frame make the overall dimensions about $5 \, \text{mm}^2$.

Higher power LEDs (typically 5 W) are also available. Whilst the principle of operation is very similar to standard LEDs the size and construction techniques are somewhat different. Presently they are used in certain countries for exit signs, automobile brake lights, striplights for path marking and emergency route identification.

However, with advances in LED technology progressing at an impressive rate, potential uses are architectural, retail display, use within aircraft and

Table 3.11 Light emitting diode output colours for semiconductor doping material

Doping material	Output colour(s)
Indium gallium nitride (InGaN)	Blue, green and white
Aluminium gallium indium phosphide (AlGaInP)	Red to amber
Aluminium gallium arsenide (AlGaAs)	Red

Table 3.12 Summary of light emitting diode characteristics

	Light emitting diodes
Commonly known as	LEDs
ILCOS code	
Wattage range (W)	0.1 per individual diode
Luminous efficacy (lm/W)	30
Colour temperature (K)	3000–8000
CRI	66–80
Average rated life (hours)	Potentially 50 000 but unproven
Burning position	Universal
Run-up time	Instantaneous
Re-strike time	Instantaneous

Note
Values in the table refer to LEDs emitting 'white' light.

displays. Slightly longer term, this source could dominate both internal and external lighting markets. The characteristics of an LED are tabulated in Table 3.12.

3.3   Control gear

3.3.1   Introduction

Discharge lighting is by far the most efficient method of illuminating any area, with the lowest associated running costs. Incandescent lamps do not require control gear, but are generally considered to be too inefficient for exterior lighting. When using discharge lamps there is a need for control gear. LEDs, depending on the type used, will either run directly on a 240 V supply or require a transformer allowing them to run at a much lower voltage.

Control gear conventionally consists of three components: a ballast, a capacitor and an ignitor. Alternatively, electronic gear may be used in some cases.

3.3.2   Role of control gear

3.3.2.1   Ballast

A ballast is a large inductor (or choke) which goes in series with the lamp and is used to limit (or choke) the current into the lamp. Without a ballast, a lamp will draw as much current as it can until it destroys itself. The ballast is manufactured to tight tolerances to correctly limit the lamp current to a level stipulated by the lamp manufacturer.

3.3.2.2 Capacitor

The capacitor, or 'power factor correction capacitor' (PFC), is used in conjunction with the ballast. A ballast is a large inductive load which will create a lagging power factor. The PFC is placed across the mains supply to correct the lagging power factor until it is near unity. Without the capacitor, the input current into the luminaire would be much higher without the benefit of increased light output.

3.3.2.3 Ignitor

The ignitor is used to initiate the discharge within the lamp. An ignitor generates pulses in the order of 4.5 kV (2 kV for 70 W HPS and below) in order to strike the lamp. Once the lamp is operational, the ignitor switches off as no more high voltage pulses are required to sustain the energy in the lamp. For types of ignitor, see Section 3.3.4.3. Some lamps have internal ignitors.

3.3.2.4 Electronic control gear

Electronic control gear is an alternative that replaces all the three components. Electronic gear is particularly popular for running fluorescent lamps and is slowly gaining popularity for the smaller discharge lamps, particularly high pressure sodium and metal halide.

3.3.3 Requirement for control gear

Not all lamps need control gear. Table 3.13 shows which lamps need which type of gear.

3.3.4 Important characteristics

3.3.4.1 Ballast

A magnetic ballast is an extremely robust device made up of a copper winding within steel laminations. Ballasts have a long lifespan due to the fact that there is little to go wrong particularly because of modern insulation systems. Ballasts are available with a number of voltage taps, the most common being 230 and 240 V, which satisfies most European applications.

A stable lamp requires the nominal lamp current. If the lamp is to be used in a country with a different voltage or frequency supply then a specific

Table 3.13 Control gear requirement of various lamp types

Lamp type	PFC	Ballast	Ignitor	Electronic option
Incandescent	No	No	No	No
Low pressure mercury (fluorescent)	Yes	Yes	Yes	Yes
Induction	No	No	No	Yes
Low pressure sodium	Yes	Yes	Yes	Yes
High pressure sodium	Yes	Yes	Yes	Yes
White high pressure sodium	Yes	Yes	Yes	Yes
High pressure mercury	Yes	Yes	No	No
Metal halide	Yes	Yes	Yes	Yes
Ceramic and quartz metal halide	Yes	Yes	Yes	Yes

Note
There are one or two exceptions to the above, the table reflects what would be considered industry standard.

ballast for these conditions will be required. If the control gear is for use in a country with a higher voltage then the ballast should have increased impedance to ensure that the current running through the lamp remains as specified. Conversely, if the control gear is to be used in a country with a higher supply frequency, the impedance must be reduced. For additional protection, ballasts are available with thermal cut-outs that switch the ballast off at a temperature below which would cause any damage.

The important parameters are temperature related, and often confused. A ballast generally has two ratings: Δt and T_w. The temperature rise of the ballast is Δt. Typical Δt ratings are between 60 and 80 °C under specific test conditions. The maximum working temperature of the winding within the ballast above which damage to the insulation may occur is T_w.

When specifying control gear for use within a luminaire, the Δt will give you an indication of the temperature increase that can be expected. It is then important to ensure that at the maximum ambient temperature the luminaire will be working at, the maximum T_w is not exceeded. As a rule of thumb for temperature against life, Table 3.14 applies.

Table 3.14 Anticipated ballast life for temperature during operation

Temperature of ballast during operation	Anticipated life (years)
T_w	10
$T_w + 10$	5
$T_w + 20$	2.5
$T_w - 10$	20
$T_w - 20$	40

3.3.4.2 Capacitor

The PFC capacitor is not a particularly active element within the lighting circuit. However, the capacitor helps to improve the power factor of the circuit ensuring that minimum electrical energy is used to power the lamp. Without one everything will function normally, the luminaire though having a very poor lagging power factor. The things to watch out for are the correct working voltage and whether the capacitor will be working within its temperature rating. Without the capacitor, the input current into the luminaire would be much higher. If the capacitor fails open circuit then the lamp will still appear to operate as if there was no capacitor. The capacitor should be checked during lamp change.

3.3.4.3 Ignitor

Electronic ignitors are available in two varieties: impulser and superimposed pulse (SIP). An impulser ignitor uses the ballast winding to generate the output voltage whilst the SIP ignitor generates the voltage by means of its own internal winding.

Impulser ignitors have to be used with a tapped ballast and generally each ballast wattage will need a dedicated ignitor. This type of ignitor can only be used with a precision wound ballast, that is a perfectly layered winding whereby the start layer cannot come in contact with the finish layer. The main advantage of using impulser ignitors is that the pulse is very wide and is capable of being used where the ballast/ignitor maybe 10 or 20 m from the lamp. Impulser ignitors where specified should have built-in timers to switch the ignitor off after 10–15 min should the lamp be missing or not have struck.

Superimposed pulse ignitors are wired in series with the ballast and do not need a ballast with a tap. Except for 70 W HPS and below, one ignitor type can generally be used up to 400 W. The 35 and 70 W HPS lamps need an ignitor that generates a lower output voltage of 2 kV rather than 4 kV for all metal halide lamps and HPS lamps above 70 W. This type of ignitor has to be mounted within 1 m of the lamp, as its output voltage pulses are very narrow. This type should also be specified with a built-in timer. SIP ignitors are more popular than impulsers, as only a few different types are required to cover many different lamp wattages and types.

Ignitors for SOX discharge lamps are simpler to understand. There is generally one ignitor per lamp wattage. These generate 700–1000 V and are connected directly across the lamp.

3.3.4.4 Electronic control gear

Electronic control gear is most popular for use with fluorescent lamps. All low wattage compact fluorescent lamps can use electronic gear. The payback is also appealing for use in commercial applications where T5 and T8 fluorescent lamps are used. By working at higher frequencies electronic gear can be smaller, lighter and give better lumens/watt. However, when using this kind of gear the ambient temperature must be considered as it is not as robust as its wound counterpart.

Electronic gear for high intensity discharge (HID) lamps is still a relatively new concept. There are a number of types available for running high pressure sodium and metal halide lamps in particular, especially at lower wattages. There are a number of different philosophies available and these should be analysed to balance out the benefits, perhaps reduced electricity consumption, improved lamp life or depreciation, against the drawbacks such as the increased cost.

3.3.5 Potential problems

If control gear is used within its temperature and voltage rating, unreliability should not be an issue. As with anything, however, compatibility should always be considered. Ballasts are usually single or dual tapped 230/240-V conditions can occur where the supply voltage is outside these voltages and the use of more special multi-tap ballasts should be considered. At the end of lamp life it is possible that a lamp will rectify. When this occurs a very large current is drawn through the ballast causing it to overheat. This is not usually a problem when lamps are changed automatically after a pre-determined period. If lamps are only changed when they fail it is possible that they have rectified, and under these circumstances it is recommended that ballasts with a thermal cut-out fitted are used. Recently, recommendations have changed and thermal cut-outs must be used with a wide variety of HID lamps.

Timed ignitors are recommended, to increase reliability. A timed ignitor will turn off within a pre-set period if a lamp is missing or has failed and will therefore reduce the stress on the ballast and associated wiring within a luminaire. This compares with a non-timed ignitor which will try to strike the lamp throughout the period power is supplied to the circuit. A timed ignitor will also switch off an end-of-life cycling lamp that may be overheating the control gear.

Discharge lamps generally need to cool down before they will restrike after being switched off. The timing depends on the type of lamp used. Typically, high pressure sodium lamps will restrike within 5 min of going out while metal halide lamps may take up to 15 min. Therefore, for example, a timed high pressure sodium ignitor will stop trying to strike a lamp after 5 min and will not try again until the circuit has been switched off and re-energized.

Care should be taken on the type of timed ignitor in the circuit if metal halide 'retrofit' lamps are used. These lamps are intended to run on standard high pressure sodium gear in place of the original lamp. It must be ensured that the timing of the ignitor suits the lamp used.

If a reputable manufacturer of control gear is chosen, and compatibility observed, there should be few issues of concern with control gear. The life of control gear may be extended by keeping it below rated temperatures.

3.4 Luminaires

3.4.1 General

A lamp is a source of light; it usually requires additional equipment to hold and operate it. A luminaire is the equipment that holds the lamp and the necessary lamp control equipment, protects the lamp, modifies the way the lamp light is redistributed to suit the application, connects to a power supply and adds other features to the product such as decoration and style.

Features to note about luminaires are

- Reflector optics are recommended
- Low profile visors (<60 mm depth) lightly curved visors are preferred to flat glass
- Louvres will reduce the percentage of light emitted
- Vandals will attempt to damage them.

3.4.2 Luminaire types

3.4.2.1 Luminaire choice

Section 3.4.2.2 illustrates common luminaire types with brief indications of their features and applications.

The luminaires shown are intended to be representative of generic types. Other luminaires may be available that fulfil the same or similar functions, in different ways, styles and sizes; consulting manufacturers' literature will show the variety and innovative approaches on offer.

Key attributes assume that certain basic requirements are met, for example durability in the external environment, resistance to water ingress.

Applications are shown for each luminaire style; these are intended as examples only. The choice of luminaire type is, of course, one for the individual designer and project. Lighting solutions can be approached in many different ways.

3.4.2.2 Typical product types

3.4.2.2.1 MAIN ROAD REFLECTOR LUMINAIRE

Used on traffic routes, motorways, town centres and industrial areas. A typical main road reflector luminaire is shown in Figure 3.3. The optical system

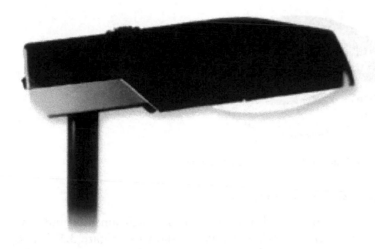

Figure 3.3 Main road reflector luminaire.

used is a twin-beam reflector (see Figure 3.36) for conventional road lighting and is optimised for luminance performance specifications. The mounting height is typically 8–15 m using an HID lamp rated up to 400 W with integral control gear.

There is often a conflict between achieving the maximum possible spacing for economic reasons, and minimising upward light and sky glow. A good compromise can often be achieved with a very low profile curved visor (typically <60 mm depth). Flat-glass visors may result in reduced spacing with no direct upward light if mounted at 0° elevation but may not necessarily produce the least overall upward light when reflected upward light is taken into account. The light distribution can usually be altered to suit different road widths by moving the relative positions of the lamp and reflector. Fixing is side entry or post mount.

3.4.2.2.2 MAIN ROAD STYLISH REFLECTOR LUMINAIRE

Used in town centres, residential, commercial, pedestrian, high amenity areas, open spaces and car parks. A typical main road stylish reflector luminaire is shown in Figure 3.4. The optical system used is a twin-beam reflector (see Figure 3.36) for conventional road lighting and is optimised for luminance or illuminance performance specifications. There are a variety of styles for decorative appearance.

A curved glazing visor allows increased spacing whereas a flat glass gives minimal upward light. Lighting performance is the same as a main road reflector luminaire but with improved daytime aesthetics. Fixing is side entry, post mount or top entry.

Figure 3.4 Main road stylish reflector luminaire.

3.4.2.2.3 HERITAGE-STYLE LUMINAIRE

Used in town centres, residential, pedestrian, high amenity and historic areas. A typical heritage-style luminaire is shown in Figure 3.5. The optical system is a twin-beam reflector (see Figure 3.36) for conventional road lighting and is optimised for luminance or illuminance performance specifications. The mounting height is typically 4–10 m using an HID lamp rated between 35 and 250 W with integral control gear. Performance is similar to a main road reflector luminaire but with significantly improved daytime

Figure 3.5 Heritage-style luminaire.

aesthetics. There are various period styles for appropriate areas and a range of sizes to suit the scale of the application. Fixing is top entry and matching style brackets and lighting columns are usually available.

3.4.2.2.4 MAIN ROAD REFRACTOR LUMINAIRE

Used on traffic routes and industrial areas. A typical main road refractor luminaire is shown in Figure 3.6. The optical system is a twin-beam refractor for conventional road lighting and is designed for luminance performance specifications. The mounting height is typically 6–12 m using an SOX discharge lamp rated up to 180 W with integral control gear.

Light control is by prisms moulded onto the visor. By adjusting the lamp in relation to the refractors, different distributions of light can be achieved. However, this type of optical system produces more upward light and less precise light control than a reflector optic. Mounting is side entry or post mount.

3.4.2.2.5 SUBSIDIARY ROAD REFLECTOR LUMINAIRE

Used in residential, pedestrian, industrial and commercial areas and car parks. A typical subsidiary road reflector luminaire is shown in Figure 3.7. The optical system is a twin-beam reflector (see Figure 3.36) for conventional road lighting and is optimised for illuminance performance specifications. The mounting height is typically 4–6 m using an HID lamp rated up to 150 W with integral control gear. Luminaires using compact fluorescent lamps have recently arrived on the market.

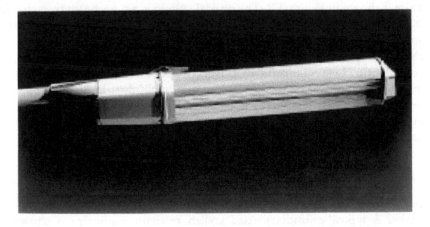

Figure 3.6 Main road refractor luminaire.

Figure 3.7 Subsidiary road reflector luminaire.

The light distribution can usually be altered to suit different road widths by moving the relative positions of the lamp and reflector. A curved glazing visor allows increased spacing whereas a flat-glass option minimises upward light. Fixing is side entry or post mount.

3.4.2.2.6 SUBSIDIARY ROAD REFRACTOR LUMINAIRE

Used in residential and pedestrian areas. A typical main road refractor luminaire is shown in Figure 3.8. The optical system is a twin-beam refractor for conventional road lighting and is optimised for illuminance performance specifications. The mounting height is typically 4–6 m using a SOX discharge or compact fluorescent lamp typically rated up to 55 W, with integral control gear. Light control is by prisms moulded onto the visor. By adjusting the lamp in relation to the refractors different distributions of light can be achieved. However, this type of optical system produces more upward light and less precise light control than a reflector optic. Fixing is side entry or post mount.

3.4.2.2.7 SUBSIDIARY ROAD STYLISH REFLECTOR LUMINAIRE

Used in residential, pedestrian, industrial, commercial, open areas and car parks. A typical subsidiary road stylish reflector luminaire is shown in Figure 3.9. The optical system is a twin-beam reflector (see Figure 3.36),

Figure 3.8 Subsidiary road refractor luminaire.

Figure 3.9 Subsidiary road stylish reflector luminaire.

variable for road and area lighting and optimised for luminance- or illuminance performance specifications. A curved glazing visor increases spacing whereas a flat-glass option gives minimal upward light. The lamp is usually an HID rated up to 150 W although CFL and induction lamps can also be used. Control gear is integral. There are a variety of styles for decorative appearance. Performance is as conventional luminaires but with particular style. Fixing is side entry, post mount or top entry.

3.4.2.2.8   POST-TOP LUMINAIRE

Used in residential, pedestrian, open areas and car parks. A typical post-top luminaire is shown in Figure 3.10. The optical system can be a bare lamp (i.e. no control), a cylindrical refractor or twin-beam reflector. Mounting height is typically 4–6 m with an HID lamp rated up to 150 W and integral control gear. There are a variety of styles for plain or decorative appearance. Glazing varies from opal to patterned to clear depending on the appearance and performance required.

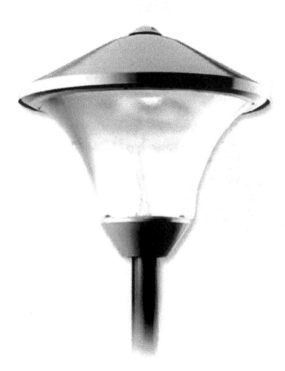

Figure 3.10 Post-top luminaire.

3.4.2.2.9   DECORATIVE POST-TOP LUMINAIRE

Used in residential, pedestrian, open areas and car parks. A typical decorative post-top luminaire is shown in Figure 3.11. Light control can be a bare lamp (i.e. no control), a cylindrical reflector or refractor, louvres or decorative enclosures. Glazing varies from opal to patterned to clear depending on the appearance and performance required. The importance of good performance should not be neglected. Mounting height is typically 4–6 m with an HID lamp rated up to 100 W, but CFL and induction lamps can also be used. There are a variety of styles for plain or decorative appearance. This category of luminaire is primarily used for appearance.

3.4.2.2.10   POST-TOP HERITAGE LUMINAIRE

Used in residential, pedestrian and historic areas and parks. A typical post-top heritage luminaire is shown in Figure 3.12. Optical control varies from cylindrical refractor to twin-beam reflector optics for road and area lighting. Mounting height is typically 4–6 m with an HID lamp rated up to 150 W,

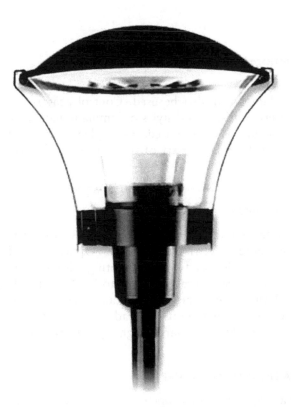

Figure 3.11 Decorative post-top luminaire.

Figure 3.12 Post-top heritage luminaire.

but CFL and induction lamps can also be used. Control gear is usually integral. There are a variety of traditional styles of luminaire most coupled with high lighting performance where required. A complementary range of lighting columns and brackets and a variety of sizes to match the scale of the application is usually available.

3.4.2.2.11   TOP-ENTRY SUBSIDIARY ROAD LUMINAIRE

Used in residential, pedestrian and historic areas and parks. A typical top entry subsidiary road luminaire is shown in Figure 3.13. Optical control is a twin-beam and variable reflector for road and area lighting. Mounting height is typically 4–6 m with an HID lamp typically rated up to 100 W, but CFL and induction lamps can also be used. This type is a small traditionally styled luminaire for top-entry mounting that maintains the traditional character of an area at the same time as giving a high standard of lighting performance.

3.4.2.2.12   INDIRECT AREA-LIGHTING LUMINAIRE

Used in pedestrian areas, squares, parks, open areas and shopping areas. A typical indirect area-lighting luminaire is shown in Figure 3.14. This type

Figure 3.13 Top-entry subsidiary road luminaire.

has an indirect lighting unit, that is the lamp is concealed from view. The light may be projected onto the reflector head from a lamp in the top of the post, or mounted in the base with a light guide up the lighting column. Both options provide a 'soft' lighting effect. Mounting height is generally 4–6 m, with a variety of reflector heads to modify the light distribution. The luminaire is primarily chosen for its style and lighting appearance. It is less efficient than direct lighting luminaires, but gives a comfortable and attractive night-time visual scene. It can also be less of a draw to vandals as there is no visor directly visible.

3.4.2.2.13 ARCHITECTURAL LUMINAIRE

Used in roads, town centres, residential and pedestrian areas, retail parks, commercial areas, open areas and car parks. A typical architectural luminaire is shown in Figure 3.15. This type has decorative features and is primarily chosen for styling, although many also have the ability to provide a high standard of lighting performance together with the required appearance. Design can be based around a standard core product to provide the style required but with standard production engineering reducing costs

Figure 3.14 Indirect area-lighting luminaire.

(a)

Figure 3.15 Architectural luminaires.

(b)

(c)

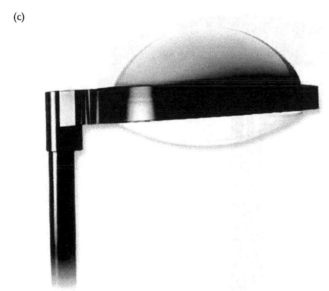

Figure 3.15 (Continued).

whilst retaining performance features. Alternatively, design can be totally customised.

3.4.2.2.14 MULTI-LAMP LUMINAIRE

Used in town centres, wide roads, junctions, roundabouts, retail parks, commercial areas, open areas and car parks. A typical multi-lamp luminaire is shown in Figure 3.16. The luminaire is post-top containing multiple optical units in one housing. The optical system has variable reflector settings to tailor light distribution to the desired area. The high light output enables a reduced number of units at a larger mounting height to cover the given area. Normally there is no direct upward light. Mounting height is typically 10–15 m with HID lamps up to 6 × 400 W.

3.4.2.2.15 CUSTOM-DESIGNED LUMINAIRE

Used in roads, town centres, residential and pedestrian areas, retail parks, commercial areas, open areas and car parks. A typical custom-designed luminaire is shown in Figure 3.17. This type has custom-designed features and is engineered largely to match the requirements of a specific project. It can incorporate standard components, and, with careful design, performance can be identical to conventional luminaires (although not for the

Figure 3.16 Multi-lamp luminaire.

Figure 3.17 Custom-designed luminaire.

type illustrated). Effects, such as lit decorative features, can be incorporated. This is generally a more expensive solution than standard product-based solutions.

3.4.2.2.16 HIGH-MAST LUMINAIRE, LARGE AREAS

Used to light large sections in industrial areas and at road interchanges, road junctions and roundabouts. A typical high-mast luminaire cluster is shown in Figure 3.18. Light distribution is adjustable to either single-, twin- beam or symmetric-beam from each luminaire so that an array of luminaires can be configured to give the desired overall distribution independent of the luminaire disposition. The luminaire is designed to be used in multiples on masts at large mounting heights, typically 18–40 m. Beam elevation may be lower than conventional road lights to minimize glare. Lamps are HID rated from 250 to 1000 W. Generally, luminaires are lowered for maintenance.

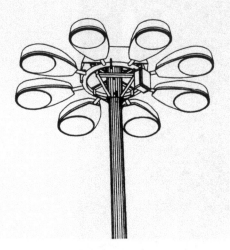

Figure 3.18 High-mast luminaire cluster.

3.4.2.2.17   CIRCULAR (HIGH POWER HIGH INTENSITY)
FLOODLIGHT PROJECTOR

Used in industrial areas, sports lighting and large areas with limited luminaire positions. A typical circular (high power high intensity) floodlight projector is shown in Figure 3.19. The optical system is a reflector designed to give a fixed beam. The beam can vary between narrow to wide dependent upon design. The narrow beam allows long distance lighting and can be aimed using the adjustable cradle mounting whilst the wide beam provides a less intense distribution over a wider area which can be aimed in the same manner as the narrow beam projector.

Usually mounted at large height on a tower, high mast or arena roof, using HID lamps rated up to 2 k W, with separate control gear. Accessories include louvres or hoods to limit glare and spill light, and a mesh glass guard.

3.4.2.2.18   CIRCULAR (MEDIUM TO LOW POWER)
FLOODLIGHT PROJECTOR

Used in building and feature effect lighting, sports lighting and for medium-sized areas with limited luminaire positions. A typical circular (medium to low power) floodlight projector is shown in Figure 3.20. The optical system is a reflector designed to give narrow to medium width beams. There is fully adjustable beam aim from a cradle mounting. Usually mounted at low to medium height on a lighting column, pole, tower or building using

Figure 3.19 Circular (high power high intensity) floodlight projector.

Figure 3.20 Circular (medium to low power) floodlight projector.

HID lamps rated from 35 to 400 W, with integral control gear. Accessories include louvres or hoods to limit glare and spill light, a mesh glass guard, and colour filters.

3.4.2.2.19   TROUGH (HIGH POWER MEDIUM INTENSITY LOW GLARE) FLOODLIGHT PROJECTOR

Used in industrial areas, sports lighting, large areas with limited luminaire positions and building effect lighting. A typical trough (high power medium intensity low glare) projector is shown in Figure 3.21. The optical system is a reflector designed to give an asymmetric narrow to wide beam in the vertical plane and a wider beam in the horizontal plane. The floodlight is mounted horizontal to minimise glare and spill light. This allows medium distance lighting with good horizontal coverage. There is adjustable beam aim from a cradle mounting.

Usually mounted at medium to large height on a tower, high mast or arena roof using HID lamps rated up to 2 kW with separate control gear. Accessories include louvres or hoods to limit glare and spill light and a mesh glass guard.

(a)

Figure 3.21 Trough (high power medium intensity low glare) floodlight projector.

(b)

Figure 3.21 (Continued).

3.4.2.2.20 TROUGH (MEDIUM TO LOW POWER AND INTENSITY) FLOODLIGHT PROJECTOR

Used in building floodlighting, and sports and feature lighting. A medium to low power and intensity floodlight projector designed to illuminate medium-sized areas with limited luminaire positions is shown in Figure 3.22. The optical system is a reflector designed to give narrow to wide beams in the vertical plane and a wider beam in the horizontal plane. There is fully adjustable beam aim from a cradle mounting.

Usually mounted on a lighting column, pole tower or building at low to medium height, using HID lamps rate from 35 to 400 W, with integral control gear. Accessories include louvres or hoods to limit glare and stray light, a mesh glass guard and colour filters.

3.4.2.2.21 GROUND FLOODLIGHTS

Used in building and feature lighting. A typical ground floodlight is shown in Figure 3.23. Comprises a watertight ground-recessed uplighting luminaire housing different lamp types and wattages and associated integral control gear as applicable. There are various reflector profiles to tailor the light

Figure 3.22 Trough (medium to low power and intensity) floodlight projector.

Figure 3.23 Ground floodlight.

distribution. Accessories include louvres to limit glare and stray light, and guards to protect the glazing and prevent contact with hot parts.

3.4.2.2.22   GROUND DECORATIVE LIGHTS

Used for route marking, boundary lighting and effect lighting. A typical ground decorative light is shown in Figure 3.24. Comprises a low power

Figure 3.24 Ground decorative light.

watertight ground-recessed luminaire, providing a visible decorative source rather than an illuminant, and contained in inconspicuous or deliberately stylish housings. The light source has low power and is tungsten halogen, CFL, LED arrays or fibre optics. Coloured light is a common feature.

3.4.2.2.23 ILLUMINATED BOLLARD

Used in footpaths and pedestrian areas, for boundary and entrance marking and to illuminate shrubbery. A typical illuminated bollard is shown in

Figure 3.25 Illuminated bollard.

Figure 3.25. Comprises a luminaire on a base unit typically up to 1 m in height housing a tungsten halogen, HID or CFL lamps rated up to 75 W and integral control gear. Exposed components that can be touched should have limited temperature. Optical control is by duplex reflector, louvre or cylindrical refractor and the unit may incorporate accessories such as shields to limit glare and the spread of light, or glazing protective bars. Use of bollards without light control is not recommended. Access is via a door, removable head or detachable base.

3.4.2.2.24 STEP LIGHT

Used to light and/or mark steps, slopes and walkways with a bounding wall. A typical step light is shown in Figure 3.26. Comprises a low level low power luminaire for recessing into steps, stairs or wall structures to provide localised lighting. The lighting is normally patchy but sufficient to adequately mark the route and show obstructions. The luminaire is waterproof, rugged and not openable without tools since it is readily accessible.

Figure 3.26 Step light.

Figure 3.27 Wall/Security luminaire.

3.4.2.2.25 WALL/SECURITY LUMINAIRE

Used in localised pathway lighting, limited-area lighting and entrance areas. A typical wall/security luminaire is shown in Figure 3.27. It is a medium low power wall-mounted luminaire suitable for surface and lighting column mounting, incorporating CFL and HID lamps rated up to 100 W and integral control gear. The optical system is usually a combined reflector and refractor optic, with the refractor integral with the glazing to help conceal the lamp and the luminaire interior. The refractor orientation gives limited light control, and as a consequence a fair amount of light pollution and glare. Construction is usually die-cast metal body and injection-moulded refractor.

3.4.3 Materials and construction

3.4.3.1 Introduction

Exterior luminaires need to meet particular environmental and operational requirements, and therefore their construction and materials must be chosen with this in mind.

At best the interior will demand protection against ingress of water, at least blown rain, and UV radiation from natural light. At worst the water will flood over the luminaire, often with considerable force. The luminaire may be subject to corrosive attack, for example seaside salt or industrial pollutants, and vibration and other physical shock, for example when exposed to wind forces on lighting columns and towers.

The luminaires also need to be relatively easy to keep clean and maintain, conditions under which maintenance must take place may not be comfortable and easily accessed.

3.4.3.2 Materials

3.4.3.2.1 CORROSION

Corrosion in metals is an important consideration. Materials must be selected that are inherently corrosion resistant such as stainless steel, aluminium, aluminium alloy, brass or bronze, or, be suitably protected, for example by painting, plating or hot-dip galvanising.

Dissimilar metals in contact can corrode through electrolytic action. Details are given in Section A.2.1.

3.4.3.2.2 UV DEGRADATION

Ultraviolet degradation occurs in plastics. The effect is to gradually break the bonds that form the long molecules. This results in loss of strength, and can lead to discoloration and surface degradation. Natural UV is a significant cause dependent upon geographical location and protection of the component to exposure. The light source will also contribute: mercury-based sources generally have a fairly high UV output but may use absorbing glass to reduce it. Operation at elevated temperature increases the speed of degradation.

The main areas to be considered together with relevant issues are outlined below. Full details are given in Section A.2.3.

3.4.3.3 Construction

3.4.3.3.1 HOUSINGS

Housings generally consist of metals such as aluminium (or alloys thereof), stainless steel, copper, copper alloys such as brass or bronze, steel, plastics such as ABS, polycarbonate, polyamide, or GRP. Each has a different combination of characteristics in terms of strength, weight, ease of construction, ease of recycling, cost, expected life, temperature resistance and corrosion resistance. A full analysis is given in Section A.2.3.

3.4.3.3.2 SUPPORTS

Luminaire supports, for example cradles for floodlights, need to be sturdy to resist vibration and movement in the wind; otherwise the light patch will move unacceptably. Adjustment should allow full movement to aim the beam and should lock positively using, for example, serrated mating surfaces or locking nuts and washers. If the luminaire requires moving for maintenance, the original aiming position should be clear to reset, for example by adjustable pre-set stop(s). Galvanised steel cradles and cast aluminium knuckle joints are often used, with aiming scales to assist setting.

3.4.3.3.3 VISORS

Glass has many advantages in terms of great durability in a wide range of environments, ability to be moulded into prisms (by pressing) and a low-cost base. In addition it can be cleaned effectively.

Plastics, either acrylic or polycarbonate, also have many advantages such as price and ease of forming into shapes. Polycarbonate is widely used in luminaries for its impact resistance.

3.4.3.3.4 FASTENINGS

External fastenings should be of corrosion-resistant materials, for example stainless steel, copper alloy or aluminium alloy, plated or (spun) galvanised. Plastic can also be used but is subject to creep and degradation.

Fastenings that are to be released for routine servicing should be easy to undo with a limited range of standard tools, or without tools. Where they are accessible to the public standard but less common tools should be used, such as recessed hexagon [Allen] head, or trihead as in lighting column doors and should be large enough to be handleable in unfriendly conditions of wet, cold, wind, and at height, and captive (retained) so they are not lost.

Fastenings should, wherever possible, be positively open or closed, not relying on the operative to adjust the setting (e.g. toggle latches, $1/4$ turn fasteners). All fasteners should be locked against vibration.

Aids for operation should be included to help the servicing operator, for example on latches pull-loops that can be operated with a gloved hand.

3.4.3.3.5 OPTICS

Reflectors   Reflectors are almost universally made from aluminium or met-allised plastic because of their high reflectivity. Advantages are a range of finishes, from specular to matt and patterned, ease of protection (by anodising), ease of forming, lightweight and cost.

The shape determines the basic light distribution and the finish, either micro or macro, modifies the shape. So a specular finish can produce a high intensity beam, with a pattern or diffusion reducing the intensity and increasing the spread. Reflectivity up to around 0.9 in total (0.85 for specular) is achieved, whilst dielectric surface layers can increase this to 0.95. The reflecting surface is usually protected by anodising, an electrochemical process that develops a controlled oxidation on the surface that is relatively transparent whilst protecting the metal against corrosion. For a reflector within a protective housing, film thickness of around 2–5 microns is used, depending on the base material. Thick films reduce specularity and reflectivity. Aluminium is widely available in pre-anodised form for fabrication, or may be post-anodised (e.g. for pressed reflectors).

Where diffuse light reflection is required, patterned aluminium can be used or white paint which may also form a general finish within the luminaire. Suitable paints that resist the operating temperatures near the lamp can have reflectances of around 0.85.

Refractors Refractors can be pressed in ordinary glass or injection moulded in acrylic or polycarbonate, with service temperature limitations being observed. Where the refractor is part of the enclosure, low expansion glass must be used. Where glass is used, fixings should not apply excessive stress otherwise it may fracture.

3.4.3.4 Seals

3.4.3.4.1 GENERAL

A prime requirement of exterior luminaires is that they should

- Resist the ingress of water since this will degrade the interior components, performance and damage the electrical integrity and
- Resist the ingress of dirt since this will degrade the lighting performance, may cause electrical damage, and be difficult to clean from the interior.

At the same time, routine access to the interior is required for servicing, usually just a lamp change. The seal(s) should therefore be designed to aid installation and servicing whilst remaining effective.

Where components are joined permanently, suitable mastic and adhesive sealants are appropriate. Where the joint is reusable, for example a removable cover or visor to gain access for initial installation or for servicing, an elastic gasket is usually fitted. This gasket may be solid rubber (silicone if high temperatures are present) sandwiched between the components and usually requires substantial force to compress to make the seal tight. Screws are often used, with ribs moulded on the gasket aiding the seal. In most cases though access needs to be obtained quickly to minimise service costs and time, and as few fastenings as possible are used. The load that can be imposed on the gasket is therefore limited, so soft gaskets of larger section are used so that the rim of the visor or cover embeds into the gasket. Close-cell sponge, a hollow section, or a combination is often used to give the softness and deformation required. The width of the component pressing into the seal influences the closing force: adding a narrow rib to a visor rim will much reduce the force required.

Materials used are commonly polychloroprene (neoprene), EPDM and silicone rubber. Silicone has the highest operational temperature (200 °C or more) and the best recovery of its original form after compression. Limiting the distance the closing component enters the gasket (to compress and deform) to about 25 per cent will assist shape recovery after opening, and aid effective resealing. If possible, bending seals should be utilised as these can have longer life as they are not compressed.

The closing component should ideally exert a uniform force on the gasket around the entire length. Hinges can interfere with this, particularly where the hinge is very near the gasket as it limits the movement to one that largely rolls the component edge on the gasket surface with limited movement to compress. Offsetting the hinge helps; using a hook rather than a hinge to retain the visor, and still allow it to hinge open, will allow all the catches to exert a uniform force and improve the seal reliability.

3.4.3.4.2   WEATHERING

Protecting the sealed area physically from the weather will aid the reliability of the seal, for example an overhanging lip that sheds water off the canopy, and that removes the force from wind blown or hosed water.

The luminaire design should try to ensure that water does not lie against the sealed joint, where it could be drawn in by capillary action or suction.

3.4.3.4.3   DRAINS

It is quite in order to allow water that may accumulate to drain out where by the nature of the product condensation is likely to form within the enclosure. Sizes of drain holes should be such as to prevent a meniscus that restricts drainage, that is 6 mm or more. Holes may be liable to obstruction from dirt. Equally, consideration should be given to dirt or insect entry. A mesh over the drain can prevent the latter.

3.4.3.5 Glands

Sealing cable entries can usually be achieved with conventional cable glands appropriate to the size and type of cable recommended. If exposed, the gland should be arranged so the cable falls away from the gland so water will not run along the cable and sit against the seal.

3.4.3.6 Pressure effects

The luminaire when hot will have a raised internal pressure; when chilled by water, rain for example, the pressure will drop below that of the surrounds. In attempting to equalise pressure by drawing in air the luminaire may also draw in any water lying against weaknesses in joints. Allowing the luminaire to 'breathe' by providing, for example, a deliberate air path from inside to outside will mitigate this. A small breather tube protected from contact with the elements can be effective.

The pressure above can also be sufficient to distort the luminaire components sufficiently to open joints. Wind pressure, and the pressure of hosed water, can have a similar effect. The strength of components, and the adhesion of bonded seals needs to be adequate.

3.4.3.7 Finishes

3.4.3.7.1 PAINT

There is a vast array of coatings applied for protection and/or cosmetic reasons. Primary requirements are adhesion and durability of protection, colour and finish. Dry powder finishes applied electrostatically and then heat cured are commonly used on metal components. Of the common ones, polyester formulations are appropriate having a good combination of properties and a wide range of RAL and BS colours and finishes. These typically cure at 200 °C so compatibility with operating temperatures needs consideration. A good pre-treatment will extend the life considerably if the paint is liable to be damaged. 'Wet' paint can be applied as a single component for heat curing at lower temperatures but provides lower durability, or as two-pack types where the additive chemically alters the paint curing it.

Epoxy-based paints tend to chalk, that is surface whiten and become powdery, when used outside.

3.4.3.7.2 ANODISING

Anodising has already been described in 'reflectors' under Section 3.4.3.3.5. Thicker films will give long protection to housings, for example 25 microns is used on architectural metalwork. The film can be coloured (dyed) with fastness depending on the colour.

3.4.3.7.3 PLATING

Usually zinc, passivated to further protect, is used on steel components. It is better if further protected (e.g. by paint).

3.4.3.7.4 ORGANIC COATING

Various proprietary coatings are available as alternatives to plating on, for example, fasteners and turned steel items, with improved protection, particularly in marine conditions. These are often applied by dipping or spraying and then thermally curing.

3.4.3.7.5 PLASTIC COATING

This is often a thick thermoplastic coat applied by dipping a hot component into the powder and then baking. Its success depends on adhesion being adequate. If the film is penetrated corrosion will spread under a poorly adhering film, as with paint.

3.4.3.7.6 GALVANISING

Hot-dip galvanising is applied to steel by immersing the item in a bath of molten zinc. A zinc–iron alloy forms on the steel, and finishes with a zinc surface. Unlike most protections this actually bonds with the steel, rather than forming a surface film. The zinc is corroded preferentially, leaving the underlying steel unharmed. Gaps in the galvanising of, typically, 3 mm will still not allow the steel to corrode. Whilst the coating is thick eventually the zinc will be used up, usually after many years. Painting will extend the life by delaying the corrosion of the zinc whilst at the same time providing a cosmetic finish.

3.4.3.7.7 METAL SPRAY

Metal, usually zinc or aluminium, is sprayed in molten form onto a prepared (shotblasted) steel surface forming overlapping platelets to protect the steel. The resultant surface is porous so a thin sealer is applied. The protection from the zinc or aluminium is sacrificial, like galvanising, but is a film not a bonded alloy. Unlike galvanising the protection can be applied only to surfaces that are accessible to the spray (so the insides of tubes are left largely unprotected).

3.4.3.8 *Control gear and wiring*

3.4.3.8.1 THERMAL CONSIDERATIONS

Components have temperature limitations that must be considered in service conditions if objective lives are to be achieved. For ballasts, the T_w rating (the temperature of the copper winding) is typically 130–140 °C. If this temperature is not exceeded, a 10-year continuous operating life is the objective. Capacitors and ignitors are rated at 90–105 °C, and electronic controllers generally 65–90 °C, the manufacturer gives a maximum case temperature T_c and usually a measuring position. Inevitably to make luminaires small and to fit large wattage lamps, operating temperatures will be near the limit. Luminaires used in unusual (e.g. enclosed) conditions or in high ambient temperatures (e.g. overseas) need to take this into consideration.

The surfaces of some components can be very hot, for example ballasts can reach 120 °C on the lamination surface. Materials such as wiring insulation need to be compatible with this. Alternatively positive wiring routes need to be provided away from the hot areas. Incoming (installer's) cable insulation will normally be PVC with a limit or 75–90 °C and will need terminations clear of the hot areas, or suitable additional protection such as glass-fibre sleeving.

3.4.3.8.2 GEAR TRAYS

Although control gear is reliable it is often convenient to mount the pre-wired circuit on a tray that is removable from the luminaire. This will often aid installation as the body can be installed and wired and tested before the gear and lamp is fitted. It makes changing easier and quicker when there is a fault, and allows many luminaires to be readily upgraded to other lamp types or ratings.

3.4.3.8.3 CONTROL GEAR LOCATION

Normally it is more convenient to house the gear within or on the luminaire. With some circuits (SIP ignitors and some electronic units) close proximity to the lamp is essential. Where operating conditions or space preclude, the gear can be placed in a separate housing, for example gear box, wall bracket or lighting column base compartment. The conditions provided must be suitable as restricted volume can result in gear overheating, placing components too close can damage electronics and excessive voltage drop can cause problems. The circuit must also be protected from the weather in the same way that it would be within a luminaire.

3.4.3.8.4 WIRING

Connections must withstand the vibration to be expected in service. Normally screw (clamp) connections will be used with stranded cable. Where solid core cable and push-terminations are used these need to be secure under vibration. Soldered stranded cable ends tend to cold-flow under pressure and become unreliable, and therefore should be avoided. Wiring routes should avoid hot components above the limits of the insulation material; otherwise additional sleeving should protect them.

Insulation should not sit against sharp or rough edges; otherwise it may be damaged by movement in service. Wiring under strain, for example incoming cables with a suspended length in a lighting column, require a cable clamp, which may be provided by a sealing gland.

High voltages produced by ignitors require suitably rated cables and connectors to be used.

3.4.3.9 Accessibility

3.4.3.9.1 INSTALLATION

Luminaires should be designed to make installation as easy as possible. Exterior conditions can be inhospitable so difficult installation can lead to poor results. Removable components such as optic or gear trays can aid accessibility to incoming wiring. These may be retained but designed to hinge clear. They can also be fitted after the basic installation has been electrically tested.

Luminaires can also be supplied totally wired and complete with lamp and an external connector for the installers supply. This has the advantage of providing a fully tested unit with minimal installation time required. The supply cable can also be factory or contractor fitted to minimise time on site.

3.4.3.9.2 MAINTENANCE

Methods to give quick access and reliable re-sealing have been dealt with above. Maintenance personnel may not be qualified to carry out full electrical servicing so a fault in a product is often more easily dealt with in the workshop. Mounting vulnerable components on removable trays, preferably with plug-and-socket–type connectors, allows on-site substitution and off-site repair.

3.4.3.10 Public protection

The public need to be protected. For example, using uncommon fixings on access components can minimise unauthorised access. Installed guards should prevent contact with hot parts, for example grills or mesh over ground-mounted luminaires. Where glass breakage would give unintentional access a secondary protection, a mesh for example, should be provided. Heavy components, or components that may cause injury if dropped from the luminaire, should have a retaining safety device. Visors that may be blown by the wind when opened for maintenance should have a captive hinge to prevent accidental detachment.

3.4.4 Light control

3.4.4.1 General

Light sources generally emit light in all directions so this light must be controlled to make the best use of it. For much external area lighting, light is emitted below the horizontal, with the distribution shaped to place it where it is wanted. Intensities above 15° below the horizontal are limited to avoid glare.

The fundamental methods used to control light can be categorised as obstruction, diffusion, refraction and reflection. It is the luminaire designer's task to use these techniques innovatively to achieve the lighting 'technical' objectives in an economic way at the same time as providing a satisfactory luminaire appearance.

3.4.4.2 Obstruction

Obstruction, or shielding, appears in most luminaires, usually to control light in the glare zone. Figure 3.28 shows this applied to a reflector. The side of the reflector is positioned to cut off the bare lamp at the appropriate angle whilst the reflector shape collects and reflects light downwards efficiently.

Louvres (Figure 3.29) fulfil a similar function. A plain white louvre both prevents a view of the bright lamp and diffuses reflected light (a) to reduce luminance. If specular material is used for the louvre, a vertical flat blade

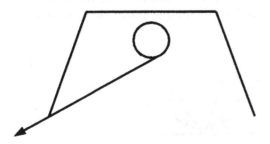

Figure 3.28 Reflector – lamp shielding.

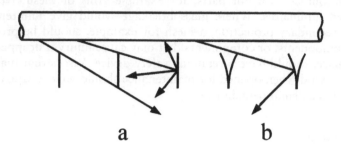

a   b

Figure 3.29 Louvres – lamp shielding.

would simply reflect a mirror image of the lamp back in the glare region; such louvres are therefore angled and often shaped to give a controlled light reflection below the glare zone (b).

3.4.4.3 Diffusion

Diffusion can be by reflection, such as from a white-painted surface. Its effect is to reduce luminance by spreading the source light over a larger area of reflector as shown in Figure 3.30.

Diffusion is also used in a similar way with transmitting materials, for example patterned or opal glass and plastic. The mechanism again is to spread the source over a larger area to make it more comfortable visually. A clear patterned diffuser breaks up the image of the source seen through it without significantly reducing the luminance of the individual elements of the image. Opal material (Figure 3.31) behaves in a similar way to a diffusing reflector by reducing the source luminance and spreading the source output over a larger area. The further the diffuser is from the source, the lower the luminance. The opal sphere is an example of minimising glare by reducing luminance, but offering no light control, at least as much light goes up as goes down. Diffusion that is less severe is used to smooth light distributions and assist visual comfort without modifying the distribution too greatly.

To shape light distribution accurately demands more controllable methods; these involve refraction to bend light, and reflection to redirect it.

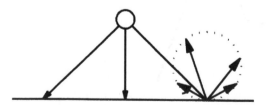

Figure 3.30 Diffuse reflection.

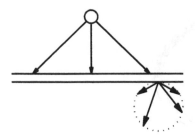

Figure 3.31 Diffuse transmission.

3.4.4.4 Refraction

Refraction requires a clear transmitting material such as glass or plastic. Light moving from air into a more refractive material will undergo a change in direction if the surface is not normal to the incident light. As it enters the material it will be refracted nearer to the normal (perpendicular) to the entry surface, and as it leaves it will be further from the normal to the exit surface.

In exterior lighting, refraction is applied in the cylindrical prismatic moulding for a vertically burning lamp as shown in Figure 3.32. The horizontal prisms and lens (A) on the outer surface refract the source light to increase the intensity broadly around 70° to the downward vertical. To increase luminaire spacing, vertical prisms on the inner surface form a bias of light to one side of the refractor often to a specific side to increase the lighting efficiency on the chosen area. If light has to be bent over a large angle, then prism losses become unacceptable and total internal reflection occurs. This effect is used by forming reflecting prisms (B) at the top of the cylindrical moulding.

Refractor control offers reasonably precise shaping of the light distribution with compact size. Injection moulding produces precise prismatic

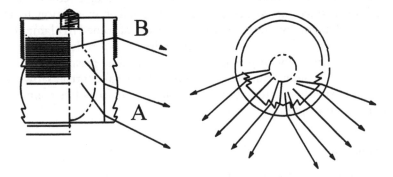

Figure 3.32 Cylindrical refractor.

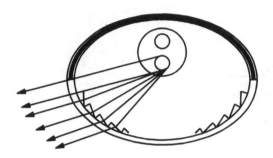

Figure 3.33 SOX luminaire refractor visor.

formations. A familiar example in road lighting is the low pressure sodium luminaire shown in Figure 3.33. The requirement is to produce two beams, up and down the road, with a beam elevation of around 70° to maximise luminance yield, above the beam intensities should fall rapidly to minimise glare. The enclosing visor has refracting prisms moulded internally. If moulded on the exterior they would form dirt traps. The prisms form images of the arc tubes over as much of the height of the enclosure as possible at the required beam elevation, a larger luminous area than the arc tubes alone is then created, but of a similar luminance, hence giving increased intensity. The source size is significantly large in relation to the prismatic visor; this limits the control that can be exercised over beam spread so upward light is not normally eliminated and glare control is less effective than with reflector optics. Additionally, long light sources like this cannot be controlled in the horizontal direction so as much light is projected behind the luminaire onto the surrounds as onto the carriageway (if the luminaire is mounted horizontally). Note, as peripheral vision is important in detecting potential hazards, this light is not all wasted.

For small sources such as HID lamps, a transparent enclosure moulded with complex arrays of prisms achieves more precise light control than for SOX lamps, since the source size is much smaller in relation to the controller. Wall luminaires and the prismatic visors of some HID lamp road luminaires are examples. Since a road luminaire with a reflecting canopy usually has a glazing visor to provide protection, adding prisms overall, or in a patch below the lamp, can provide useful additional control. However, more light is normally emitted above the beam and glare is less controlled than from reflector optics.

3.4.4.5 Reflection

Reflection returns light from a surface which can vary from a diffuse material through to a specular one that produces mirror images of the light

source. The angle(s) at which the reflective material is oriented in relation
to the light source determines the resultant light distribution. Aluminium
is the almost universal material as it offers a high reflectance – typically
0.85–0.9 when brightened and anodised, but with higher values up to 0.95
possible when carefully controlled thin films are deposited on the surface.

Aluminium is readily formed by bending, spinning and deep drawing,
and may form the structure as well as the surface of the reflector, or it may
be vacuum-deposited onto another material, such as an injection-moulded
plastic component that reproduces the required shape.

A plane specular surface reflects a ray of light making an angle α with
the normal to the surface at an equal angle α on the opposite side of the
normal; rotating the surface 1° will therefore move the reflected ray 2°.
A familiar reflector is the parabola shown in Figure 3.34. The shape is
such as to reflect all rays collected from the source into a beam parallel to
the axis. Figure 3.34(a) shows a long focal length parabola, but collecting
relatively little source light compared to the more practical one shown in
Figure 3.34(b). Light not incident on the reflector is, of course, uncontrolled
and its effect is dependent on the characteristics of the light source only. In
reality, many light sources have significant size that introduces divergence
of the beam from the axis, more so in (b) than (a) since the light source is
closer to the reflector. Introducing a refractor to fill the uncontrolled zone
(c) will increase the capacity for light control.

Reflectors based on these shapes, with the profile rotated around the
axis, are used in floodlights where relatively narrow beams for long distance
lighting or spotlighting are required. Similar profiles extruded into a trough
give a narrow beam in one plane and a fanned beam in the plane at 90°,
more appropriate to area lighting with shorter distance throws.

Twin-beam reflector luminaires are used for road lighting applications
with HID lamps. The photometric requirements are the same, two beams of
relatively high intensity projected up and down the length of the road. When
at about 65–70° above the downward vertical these efficiently produce a
good uniformity and level of road-surface luminance at spacings of up to
around 5 times mounting height, with acceptable glare control. Typically

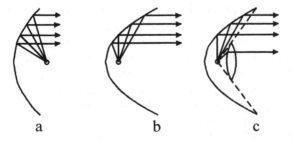

Figure 3.34 Reflectors – increasing light collection.

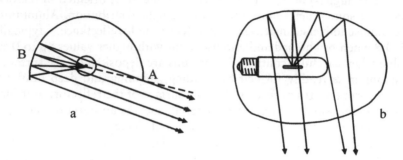

Figure 3.35 Typical one-piece road reflector.

such reflectors are formed in two planes as in Figure 3.35. Figure 3.35(a) shows a vertical section; the lower edge A (in the observer's direction) is positioned so that it prevents the arc tube being seen above about 75°, to prevent unacceptable glare. The reflecting section B is shaped to redirect source light into the beam. The width between the opposite reflector sections (i.e. the reflector mouth) must allow this light to exit the reflector without being obstructed by the opposite side, and it should also exit largely below the lamp to minimise obstruction losses.

Further beam shaping is done by the shape of the reflectors in plan (Figure 3.35(b)). By curving the reflectors in this plane, further focussing can be made into the main beams to increase the intensity, and the two beams can also be angled in towards the road (giving a beam toe). This toe can be altered to suit different widths of road by adjusting the horizontal position of the lamp in relation to the reflector.

Such reflectors collect some 60 per cent or so of the lamp flux, the remainder being emitted uncontrolled through the mouth. Depending upon the characteristics of the reflecting system and the application it may be advantageous to add diffusing or prismatic control onto the transparent visor that covers the reflector mouth, as noted above.

The kinds of intensity distributions produced in plan (azimuth distribution through the peak) are illustrated in Figure 3.36. It is not advisable to interpret too much from the proportions of these curves, but they illustrate that the majority of the light is directed onto the carriageway and near pavement. The three distributions represent different toes produced by moving the lamp. Such twin-beam reflectors are extremely good at lighting roadways, with the relatively small range adjustment to the distribution suited to the normal carriageway/pavement geometries. However, they are not normally capable of projecting much light well behind the lighting column, as this is generally not required.

Many exterior lighting applications require more varied light distributions. Whilst a conventional traffic route is assumed to require light for

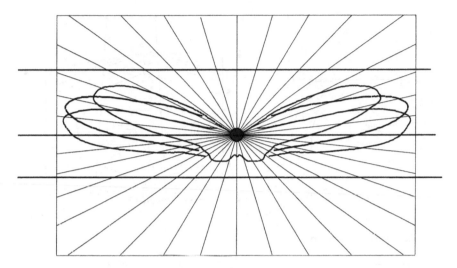

Figure 3.36 Typical twin-beam intensity distributions (in plan) (from Figure 3.35).

the drivers and the immediate surrounds, the same road running through a town will often have wide surrounding areas for pedestrians and shoppers. A residential road or homezone will have verges and house frontages where light is required for security and to create a pleasant visual scene; car parks require light over large open spaces, and shopping precincts need light specifically tailored to the environment of the pedestrian.

Such areas are of widely different shapes and if twin-beam optics are to be used then the luminaire must have a selection of optics in order that its external appearance remains constant. In such a situation an effective solution can be from variable reflector luminaires. Figure 3.37 illustrates the principle. The vertical profile of the reflector elements is smilar to the twin-beam, producing a beam at an elevation of 65–70°. The reflector is not, however, of fixed construction, but consists of four separate elements around the source, each element producing a beam. The elements are pivoted and can be moved in the horizontal plane so that their individual beams are independently aimed to give a wide range of combined light distributions.

Figure 3.38 illustrates three typical intensity distributions in comparison with Figure 3.36. In (a) the reflectors are adjusted so their beams are as far apart as possible; light is spread all around the pole position and would be used for large areas such as car parks and plazas. In (b) the reflectors are aligned to give an axial distribution, for stretched pole spacing on more linear installations; (c) is more like a twin-beam but with the advantage of the ability to control light behind the lighting column, as well as on the carriageway side. Such optical systems have the ability, by tailoring the light

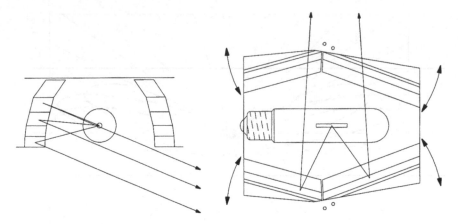

Figure 3.37 Variable reflector.

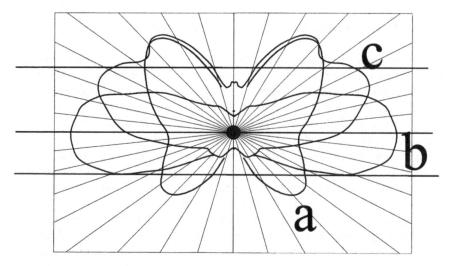

Figure 3.38 Typical range of intensity distributions (in plan) (from Figure 3.38).

distribution to better match the area being lit, to reduce the number and/or power of the luminaires required by increasing the utilisation of lamp flux.

3.4.4.6 Choice of optical system

At the technical level, the choice will depend upon the efficiency with which the competing systems fulfil the lighting requirements, i.e. the luminance and/or illuminance on the chosen planes, uniformity and freedom from glare. Light output ratio is just a crude measure; the distribution pattern of the light is of greater importance so flux utilisation is a better measure. The spacing and height form a key economic element so really comparing calculated installations of competing optics is the better way to make meaningful judgements.

But technical performance is only one aspect. The lit appearance of the luminaire is important in many installations, especially those in populated environments; the type of optical system will be important in creating the visual effect.

Luminaires (and lighting columns and brackets) in exterior lighting installations are usually a very visible part of the environment by day, by necessity or design. The daytime appearance of the luminaire, including the visible optic, needs consideration particularly in the many areas where more stylish lighting is used.

3.4.5 Performance and standards

3.4.5.1 Safety standards

3.4.5.1.1 GENERAL

The European standard embracing luminaires is EN 60598, issued in the UK as BS EN 60598. The standard is a safety standard, covering electrical, thermal and mechanical safety. It is not intended as a performance standard, although some levels of safety can also lead to improved performance (dust ingress for example). Compliance with the standard will meet the requirements of the Low Voltage Directive (LVD).

Compliance with the standard can be demonstrated in the following ways.

- By independent testing by an approved laboratory. Of itself this will be a type test result applicable to the sample(s) tested.
- When used by a manufacturer with ISO 9000 registration and with regular factory audits to monitor consistency of manufacture (quality), then a safety mark can be applied – for example the familiar Kitemark of BSI or ENEC mark. Such approvals are recognised throughout the EEC, and many other countries.
- Self-testing using test procedures that meet the requirements of the standard will allow manufacturers to demonstrate their claimed compliance with all or appropriate requirements of the standard.

3.4.5.1.2 ELECTROMAGNETIC COMPATIBILITY (EMC)

As electrical equipment, luminaires must meet the requirements of the Electromagnetic Compatibility Regulations. These require that the equipment shall not produce interference that prevents the correct operation of other electrical equipment, and is not itself affected by interference from other equipment.

3.4.5.1.3 CE MARKING

Luminaires offered for sale in the EC must meet CE Marking regulations. In essence a product must comply with both the LVD and EMC legislation. It will do so if it meets BS EN 60598 (LVD requirements met) and EMC standards.

The luminaire must display the CE mark to comply. The compliance may at its simplest be a declaration by the manufacturer (self-declaration). Enforcement is currently the responsibility of Trading Standards Officers and in the event of a challenge the manufacturers must be in a position to substantiate their claim of compliance. This can be on the basis of self-testing and technical construction file, or by independent certification.

3.4.5.2   Testing and performance

3.4.5.2.1   GENERAL

There are a number of key performance and safety tests to show how the standard relates to luminaires in practice.

3.4.5.2.2   INGRESS

This relates to the ability of a luminaire to withstand the ingress of solid objects, dust (dirt) and moisture. Details are given in Section A.2.2.

The dust tests indicate the protection against degradation due to dirt entering the luminaire and reducing light output in both quantity and distribution. Maintenance factors in the lighting design standards take account of this in recommending maintenance (light loss) factors. A high ingress-protection (IP) rating is of value since cleaning the inside of a luminaire effectively is usually difficult and relatively expensive.

3.4.5.2.3   THERMAL

The materials and components used in the construction of luminaires have maximum operating temperatures. The luminaire design needs to ensure that these are not exceeded under operating conditions, which includes fault conditions such as a rectifying lamp. Tests will include a supply voltage at the extreme tolerance to ensure that worst conditions are represented, and in an enclosure that eliminates air movement that would otherwise provide some cooling effect. The luminaire is operated in the most onerous operating position with the worst combination of components.

Component and material temperatures are normally measured using thermocouples. The winding (coil) temperatures of magnetic (wirewound) ballasts and transformers are evaluated indirectly by measuring their change in resistance from cold to operating condition – this allows their mean temperature to be accurately calculated.

The values obtained are assessed against the manufacturer's data and the limits are published in BS EN 60598. A concession for exterior luminaires exposed to free air movement is that 10 °C can be deducted to allow for the cooling effect produced.

3.4.5.2.4 WIND LOAD

Luminaires used outside, for example road luminaires and most floodlights, will normally be exposed to wind forces. This loading is considerable and imposes stress on both the luminaire construction and its fixings. To test their ability to withstand this loading, the luminaire is secured by its normal fixing method but in a position that allows a load to be evenly distributed over the side that would be exposed to the wind; normally this requires the luminaire to be mounted with its side elevation in the horizontal plane, when the load can be applied by, e.g. sandbags.

The load for road lights is specified as $1.5 \, kN/m^2$ (of maximum area presented to the wind) for mounting heights of up to 8 m, $2.0 \, kN/m^2$ for 8 m to less than 15 m and $2.4 \, kN/m^2$ at 15 m and over. The test requires no damage, and no residual permanent deformation of more than 1°.

For floodlights used above ground level the test load is set at $2.4 \, kN/m^2$ for all heights. The limit on permanent set after the test is the same 1°.

3.4.5.2.5 LIGHTING PERFORMANCE

This is not a safety issue unlike the above, but a measure of the effectiveness of the luminaire in fulfilling its prime function: distributing light effectively and economically to achieve the lighting objectives of the installation designer.

Performance can be assessed in a number of ways. Measures considered will include the efficient use of lamp flux:

- For example light output ratio – a crude gauge of efficiency in that it ignores light distribution
- Utilisation factor deals with incident flux on the area under consideration but again not its distribution
- ULR – a possible measure of light pollution if upward light is regarded as wasted (in some instances it is desirable if controlled to minimise spill light).

However, the performance of the whole installation will decide the effectiveness of the individual luminaire(s) chosen. This will take account of the efficient use of light to illuminate the task (whatever the plane of consideration), and whether illuminance or luminance is the measure, glare (e.g. threshold increment), and spill (wasted) light whether upward pollution or obtrusive light falling outside the area under consideration (information about performance measurement is given in Section A.2.4.).

3.5 Switching controls

3.5.1 Introduction

Exterior lighting by its very nature requires some form of control to switch it on or off when required. Burning exterior lighting 24 hours per day is wasteful of energy and the earth's finite resources.

Therefore, when planning an exterior lighting system, attention should be given to the control and operation of the lighting. It is generally more cost-effective to install well-conceived and considered switching arrangements with a new or refurbished lighting system than to retrofit them at a later date.

In addition the use of suitable switching arrangements will reduce energy and maintenance cost by providing only the level of illumination required for the task and help to reduce unnecessary light being sent in to the night sky.

3.5.2 Considerations

The controls for an exterior lighting system should allow

- The system to be switched on and off as required
- Individual lamps or groups of lamps to be switched on and off as required
- Lighting levels to be varied to match the level and complexity of the task.

However, the controls should be simple to understand and operate and accessible otherwise they will not be used.

3.5.3 Choice of control

3.5.3.1 Manual control

The simplest control is the common switch which either directly or through a contactor in a large installation switches on or off the lights when operated. This simple control is easy to operate and if placed in a sensible location may be operated to suit the particular requirements. It is very good for installations which have no set operational pattern and need maximum flexibility. However, it is prone to error by people switching on or off early or late endangering life or being wasteful of energy. In large installations several switches may be required to switch different groups of lamps allowing more flexibility and control over the lighting.

3.5.3.2 Time switch control

If the task or operation is consistent and performed at regular times then the simple switch can be replaced by a time switch which allows the lighting to be switched on or off at predetermined times. However, as daylight varies both in intensity and time throughout the year, consideration should be given to the use of a time switch fitted with a solar dial to allow the automatic correction for dusk and dawn. Solar dial time switches can be adjusted to allow an offset from dusk and dawn, that is the lights can be set to switch a predetermined time before or after dusk or dawn to allow for the run-up of the lamp to full brightness or for some other similar reason. They can also be fitted with secondary switching to allow the lights to be switched off at a predetermined time, say midnight, and then if required switched back on early in the morning before dawn. This arrangement is very useful for a factory or some other organisation which works fixed times and does not require all the external lighting to be lit all night but wishes the lighting to be operational during darkness when staff are on site.

The mechanical time switch has generally been replaced by fully electronic time switches which have a profile of the times of dusk and dawn throughout the year stored in their memory. Electronic time switches can be adjusted in the same manner as the mechanical time switch and many can be adjusted to give different switch on and off periods to suit different work patterns throughout the week. For instance, they can be set to switch the lighting on only on weekdays if the plant does not operate at the weekend. However, a simple and accessible override should be provided to allow the lighting to be manually switched to take account of unusual work patterns and operations.

3.5.3.3 Photoelectric cell control

Where lighting is required to be operated all night long for safety or security, a more practical control is the photoelectric cell control. This device measures the level of light at the installation and at a predetermined lux value switches the lighting on or off thus taking account of the variation in natural lighting levels due to cloud cover and other weather patterns. Photoelectric cell controls either can be factory preset to the desired switching level or can be supplied to allow the end-user to adjust the levels to suit their own specific requirements.

The simplest photoelectric cell is the thermal photocell in which a cadmium sulphide cell operates a thermal relay to switch the load. The thermal relay provides a built-in delay to take account of spurious switching due to the passing of a dark cloud or light falling on to the cell. Thermal photoelectric cells have a positive switching ratio, that is if the cell switches on at 100 lux it will switch off at 200 lux or more. Whilst cheap and effective, thermal cells are difficult to calibrate and tend to drift making them inaccurate and wasteful of energy.

In the hybrid type the cadmium sulphide cell is retained with the output being amplified to switch a thermal relay. This arrangement allows more accurate control and setting with reduced drift. However, consistency of switching levels cannot be guaranteed between different batches due to the relatively crude method of calibration used.

Modern photoelectric cells are fully electronic using photo diodes or photo transistors for light detection. The output from these is then amplified in an electronic circuit which allows for adjustment and the necessary time delay. Switching is then achieved using a triac, electromagnetic relay, reed switch or a combination of these. Discharge lamps need time (up to 5 min) to run up to full output, therefore lamps need to be switched on before they are required and before natural daylight has dropped below the level of lighting to be provided. However, to allow lamps to be switched off at the same high lighting level is wasteful of energy, therefore modern photocells generally have a negative switch on/off ratio generally of 1:0.5.

Modern electronic photoelectric cells are very accurate and can be factory calibrated to any required setting with most being calibrated to switch on at 70 lux and off at 35 lux to minimise the burning hours of the apparatus.

Photoelectric cells can be used in the same manner as time switches to control all the lighting or to control discrete groups of lamps. In general, photocells cannot switch lamps off at a predetermined time relying fully on light measurement. However, there are a number of hybrid fully electronic controllers on the market which use a photocell as a detector to switch the lights on and off and also have an inbuilt electronic algorithm which allows the controller to determine time and therefore switch the lighting off at a predetermined time rather than waiting for the correct lighting level.

3.5.3.4 Remote switching

The units previously described are either used to group switch a lighting system or part of a system at a main controller or are individually mounted in each lighting unit and operate independently of the others in the system.

Developments in modern electronics have allowed systems to be developed which can control individual lighting points from one central point using radio waves, microwaves or electronic signals transmitted along the mains power cables (mains-borne signalling) or separate pilot cables. These systems can switch individual lamps by electronically addressing each lamp separately or they can switch all lamps on simultaneously. Further developments of these systems allow data from individual lamps to be returned to the master controller to inform if the lamp is lit or not and, with the addition of extra sensors, other information on the state of the lighting can be detected and reported. These systems are still in their infancy and can suffer from electromagnetic interference from other electronic apparatus. In particular, systems that use mains-borne signalling can be heavily affected by spikes and other harmonics on the mains and need complicated filters to reduce spurious switching and distorted signals. Their use is further complicated by the ownership of the distribution network and the possibility of additional payments being required by the electricity companies for the additional use of their distribution network for the transmission of data.

3.5.3.5 Variable lighting level control

All of the above systems basically switch lights on or off with variations in lighting levels being achieved by the individual switching of individual or groups of luminaires or lamps. Unless the lighting system is carefully designed this can lead to poor levels of uniformity or dark spots. Uniformity can be maintained by the switching of individual lamps in multi-lamp luminaires, a practice commonly used in Europe for some road lighting systems. The lamp left operating will still operate at its maximum efficacy; however, the overall efficacy of both lamps when operating may be lower than that of a single lamp with the same output. For instance to achieve an output of 32 000 lm a single 250 W SON/T lamp with an efficacy of 128 lm/W can be used, alternatively two 150 W SON/T lamps with a lamp efficacy of 110 lm/W each can be used.

Some modern electronic control gear now allows individual lamps to be dimmed from full brightness down to approximately 50 per cent. Dependent upon the type of control gear and controllers used, dimming can be achieved allowing any level of lighting to be set from the highest to the lowest; alternatively preset levels can be switched as and when required. Electronic dimming allows full lighting to be provided when required for operational reasons whilst allowing reduced lighting levels to be provided for security and general movement around the site without loss of uniformity. Because the lamp is already operational, the time required to bring the light up to full brightness is very short in comparison to a cold lamp.

Activation of the dimming can be done by individual control units such as a time switch fitted to each lighting point or group of lighting points. This is particularly applicable to lighting points using preset levels of dimming. However, if continuous dimming over the full range of lighting levels is required then a remote switching system is necessary to allow total control over each individual unit. This type of system can be further enhanced by the use of additional sensors for external conditions, for example on a road lighting system vehicle detectors could be used to count the number of vehicles travelling on the road and adjust the lighting accordingly.

In a well-designed and specified lighting system, dimming of discharge lamps will reduce the overall energy consumed by the lighting system. However, a discharge lamp operating at 50 per cent of full brightness will consume approximately 60 per cent of the energy consumed at full brightness. The cost of the reduction in energy consumed has to be offset against the additional cost of the control gear and the control system required to activate the dimming and its future maintenance. Dimming of discharge lamps in exterior lighting systems may not be cost-effective at the time of going to press, however, it does help to achieve saving in the consumption of energy, reducing pollution and saving the world's finite resources.

3.5.3.6 Passive infrared detector (PIR) control

Passive infrared detectors are commonly used on domestic security lighting to detect the presence of a person and switch on the light to illuminate them. Such units should be adjusted to ensure that they only switch when the person is on or approaching the property and should be set to switch off at the shortest time possible. This will reduce energy consumption and more importantly annoyance to neighbours. PIRs can also be used in a similar manner for lighting on larger systems, however, their main drawback is that they need a light source that can be instantly switched on or off and are therefore mainly used with tungsten halogen lamps or in some instances compact fluorescent lamps. This can restrict their usefulness due to the relatively low light outputs of the compact fluorescent lamps or the high energy cost of the tungsten halogen lamps.

3.5.4 Recommendations

It is recommended that all exterior lighting systems be fitted with suitable controls to switch and/or vary the lighting levels as necessary for the particular task being carried out (Table 3.15).

Table 3.15 Recommended switching levels

Application	Switching level (lux)	
	On	Off
Road lighting	70	35
Exterior car parking lighting	70	35
Exterior industrial lighting	100	50

3.6 Supports

3.6.1 Introduction

All luminaires have to be held in a secure position by means of a structural support to ensure the lighting distribution required is maintained during the life of the installation. Due consideration has to be given to the security and maintenance of such supports which will vary from simple brackets, for building mounting, to major structures such as 60-m-high masts.

Each type of support has its own particular requirements and choice of materials. In many cases there will be national, European or international standards that should be complied with (see Bibliography).

The location and use will have a significant effect on the choice of structure, whether it is a prestigious site demanding a unique or individual approach where price is a relatively minor concern, or a more mundane, high volume installation using standard, low cost products.

3.6.2 Building mountings

3.6.2.1 General

Building mounting brackets fall into two main groups: cantilever arms for fixing to vertical surfaces and special attachments to overcome unusual fixing arrangements. Steel or cast aluminium are the most common materials and range from a simple cantilever tube mounted on a flange plate with bolt fixings to a more ornate bracket which may be specifically designed to compliment the design of a particular luminaire or lighting cluster.

For public areas in England and Wales, building mountings are not popular due to the protracted process of obtaining the necessary wayleaves. However, in Scotland the Roads Authority can serve notice under Section 35(5) of the Roads (Scotland) Act and install 28 days later if the owner has not successfully appealed to the Sheriff.

3.6.2.2 Design considerations

The structural considerations must allow for the vertical mass that is to be supported and the horizontal wind loading acting on the total projected area. In both cases a moment, or prising force, will be generated putting the fixing bolts into tension. Not only do the bolt fixings have to be designed to be capable of supporting these loads, but the wall or other structure must be capable of supporting the loads without damage or undue deflection.

For new buildings it must be established that mobile access platforms can access the wall mounting after the building and surrounding obstacles (e.g. bollards, trees, other buildings) have been completed. The other design consideration is the route for the electrical supply cable and the feasibility of future maintenance to it.

The wind loads may be derived by reference to BS EN 40. (For further detail, see Clause 3.6.4.3.)

3.6.3 Illuminated bollards

3.6.3.1 General

Bollards are frequently used with integral lighting, either for illumination or for decorative effect. A great variety of designs and materials are in use.

The main requirements are safety and resistance to vandalism due to the easy access at a low height, but due consideration must be given to maintenance and the practical aspects of authorised access.

In the UK, bollards should comply with the requirements of BS EN 60598 'Luminaires'.

3.6.4 Lighting columns

3.6.4.1 General

Lighting columns are produced in a great variety of designs and materials. Steel, particularly tubular steel, has increased in popularity in recent years in the UK for standard street lighting columns and has taken over from concrete which was popular when steel was a scarce resource during the 1940s and 1950s. Concrete lighting columns are still very popular in other European countries such as Spain and Portugal, while aluminium is more frequently used in the Netherlands. The choice of materials is very dependent upon the local economic and political climate together with the perception of the collision risk in the country in question.

In addition to the material differences, there are several different configuration requirements for mounting the lighting fittings at the top of the lighting column, which are briefly described below.

All lighting columns whether conventional street lighting columns or special ornate or prestigious types should be designed to meet the structural requirements of the national or international standards. In the UK and Europe the specific standard would be BS EN 40.

Conventional street lighting columns in the UK and Europe have to comply with all the requirements of this standard. A more detailed review of lighting column history, types and materials together with advice on inspection regimes is given in ILE Technical Report No. 22.

3.6.4.2 Geometric design

Lighting columns are frequently sited at the rear of verges or footways where they are less likely to be damaged by vehicle impact, and are provided with a bracket projecting the luminaire towards the carriageway. With improved light control from luminaires there is a trend towards shorter projections. The bracket will generally provide a near horizontal luminaire fixing for side-entry luminaires, but top-entry fixings are still used occasionally.

It is important to avoid the use of horizontal brackets and luminaire fixings, as there is a risk of giving the appearance of drooping luminaires due to the manufacturing tolerances and deflections. In addition there is increased risk of water penetration into the luminaire. A small angular upward tilt is, therefore, generally advisable.

Brackets do not always support a single luminaire, even a single-arm bracket is occasionally required to support two side-entry luminaires in parallel, and double-arm brackets are common for the central reserve of dual carriageways. At intersections, multiple brackets of three, four or more brackets are sometimes required.

Post-top lighting columns are also a regular requirement either with a vertically mounted (post-top) luminaire for area illumination or a horizontally mounted (post-mount) luminaire for footpath or carriageway illumination.

A base compartment is provided to house the electrical connection and isolating equipment mounted on an equipment backboard, normally at 600 mm above ground level, but at an absolute minimum height of 300 mm. The door closing this compartment is only regarded as a first line of defence to protect the public from contact with live electrical connections, and the internal equipment has to comply with the electrical safety legislation providing very low risk to the public even when a door is missing.

For areas subject to severe vandalism the door can be placed at a greater height, or even near the top of the lighting column, although this necessarily increases the shaft diameter, and, if vehicle access is not available, restricts the mounting height to that maintainable by ladders, where these can be used with due regard to Health and Safety requirements. Care should be taken to ensure that lighting column of this design is situated in areas where they will not be prone to damage from vehicular impact which, in a serious impact, could trap or damage the electricity supply cable potentially energising the lighting column.

The last main feature of lighting column types is the method of installation in the ground or connection to a lower structure by means of connecting bolts. These are referred to as 'planted columns', including a root for direct burial with or without a concrete surround, and 'flanged columns' where the flange plate provides the structural connection for the embedded foundation bolts.

3.6.4.3   Structural design

Lighting columns not only have to support the weight of the luminaires, fittings, brackets and attachments, but also the far more significant loading from the wind pressure acting on the projected area of the whole structure. As a relatively flexible cantilever the column loadings have to take into account the deflections and dynamic nature of wind gust loading. Wind loads have to be based on statistical data for the area and the lighting column designed for the loading that is unlikely to be exceeded in a given period of years. In Europe this is generally taken as 25 years and is known as the wind return period. This is not the same as the expected life of the lighting column, which should be significantly longer for a correctly designed and maintained installation. The design, loading and verification is specified in BS EN 40, although other standards may be applicable outwith Europe. The wind pressure varies with the location, the height above local ground level, the height of the location above sea level, the distance from the coastline and the degree of obstructions or sheltering in the locality. Unlike the design of an individual structure some simplification is needed to cover the area over which lighting columns are to be installed, and National Guidance is generally available. In the UK the wind pressure increases from the southeast to the northwest and factors are tabulated against regional authority areas to simplify the determination of the wind loading in an area.

Banners on lighting columns result in even higher loadings as they are generally larger and are mounted at greater height (see Figure 3.40). Catenary cables or festive decorations can also impose serious loads on lighting columns and can cause collapse.

Flower baskets or other loose attachments will also increase wind loading, but conversely may have the advantage of providing some damping against vibration and deflection. If flower baskets are required there is considerable structural benefit in using baskets hanging from a short projection arm by means of chains. The movement of the basket under wind load will reduce the loading on the lighting column and provide considerable damping to deflections of the lighting column under high winds. One or two balanced baskets close to the lighting column shaft are preferable to fixed baskets clamped around the lighting column shaft, where the loading will be much higher and damping non-existent, problems with corrosion of the lighting column shaft can also arise.

Even a short single-arm bracket can be very effective in providing damping against gust loading as compared with the post-top or post-mount luminaire which has no means of limiting the deflection and vibration of the lighting column shaft.

It is, however, essential that all lighting columns are checked for structural adequacy before attachments are added, and it should be noted that the use of baskets on chains will not result in a lighter new lighting column, or a

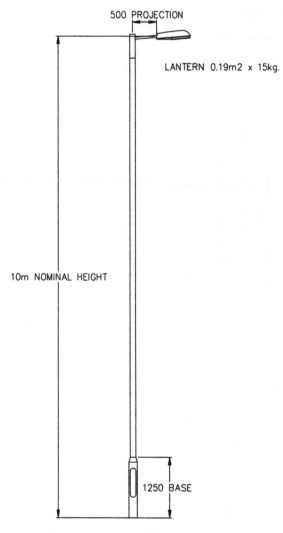

Figure 3.39 Typical 10 m lighting column (shaft 114 mm diameter 3.6 mm thick; base 168 mm diameter 5.0 mm thick; K = 2.2).

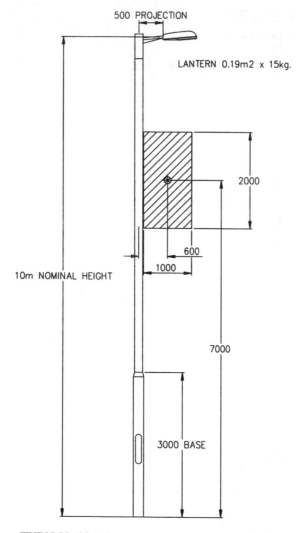

TYPICAL 10m COLUMN WITH BANNER k=2.2

SHAFT 168mm ⌀ 5.0mm THICK.
BASE   219mm ⌀ 6.3mm THICK.

⊗ SIGN CENTRE OF WIND ACTION

All dimensions are in millimetres

Figure 3.40 Typical 10 m lighting column with banner (shaft 168 mm diameter 5.0 mm thick; base 219 mm diameter 6.3 mm thick; $K = 2.2$).

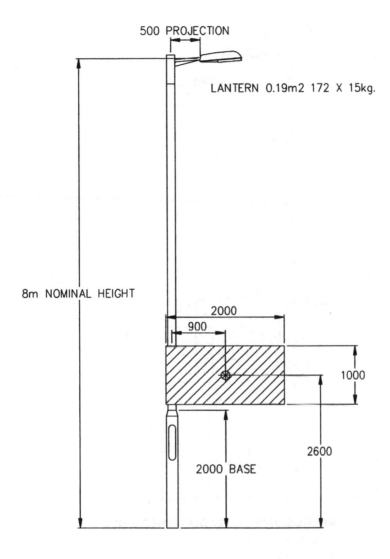

500 PROJECTION

LANTERN 0.19m2 172 X 15kg.

8m NOMINAL HEIGHT

2000

900

1000

2600

2000 BASE

TYPICAL 8m COLUMN WITH SIGN k=2.2

SHAFT 140mm ⌀ 3.0mm THICK.
BASE 194mm ⌀ 5.0mm THICK.

⊗ SIGN CENTRE OF WIND ACTION

All dimensions are in millimetres

Figure 3.41 Typical 10 m lighting column with sign and no sign support (shaft 140 mm diameter 3.0 mm thick; base 194 mm diameter 5.0 mm thick; $K = 2.2$).

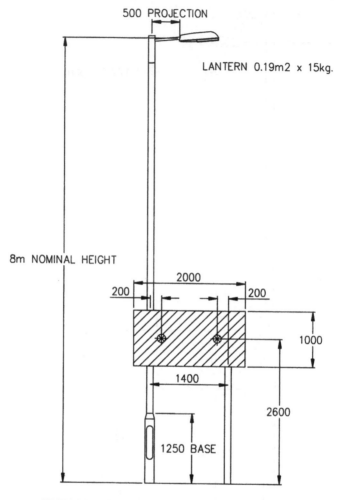

TYPICAL 8m COLUMN WITH SUPPORT k=2.2

SHAFT 114mm ⌀ 3.2mm THICK.
BASE 168mm ⌀ 4.5mm THICK.
SUPPORT LEG 114mm ⌀ 3.0mm THICK.

⊗ SIGN CENTRE OF WIND ACTION

All dimensions are in millimetres

Figure 3.42 Lighting column with sign plus sign support (sign support 114 diameter 3.0 mm thick; shaft 114 mm diameter 3.2 mm thick; base 168 mm diameter 4.5 mm thick; $K = 2.2$).

greater carrying capacity of an existing lighting column, but will improve the structural performance during the service life by reducing actual lighting column deflection.

Although designs will vary considerably, the effect of larger luminaires, brackets and signs will have a significant effect on the required strength of a lighting column. In order to give some appreciation of these effects, if a typical 10 m lighting column (see Figure 3.39) is considered with two sizes of luminaire, three sizes of sign and four bracket projections, the typical minimum thickness of the base of the lighting column can be calculated. The actual calculated values are tabulated (Table 3.16) and shown in graphical form (Figure 3.43), although in production the next available standard thickness of tube would be used. This data gives some guidance to the significance of corrosion on existing columns.

The majority of lighting column designs are verified by calculation but for specialist designs testing of the whole lighting column or particular details, such as the door openings may be required.

3.6.5 Poles

3.6.5.1 General

The term 'poles' is generally used to imply a lighting column that carries one or more luminaires for sports, architectural or floodlighting. The luminaires are therefore supported in as tight a cluster as possible at the top of the pole to limit visual intrusion. While a bracket may be required, it will supply the necessary fixing points at the minimum distance required between

Table 3.16 Simplified guidance for minimum thickness for a particular 10 m stepped tubular lighting column

Ref. no.	Luminaire		Sign (centre)			Minimum thickness in the door area for given single arm bracket projection (m)			
	Area (sq m)	Mass (kg)	Area (sq m)	Offset (mm)	Height (m)	0.5	1.0	1.5	2.0
1	0.12	15	0.0	0	0	3.6	3.9	4.3	4.8
2	0.19	15	0.0	0	0	3.9	4.3	4.8	5.4
3	0.12	15	0.3	300	2.5	4.2	4.5	4.9	5.4
4	0.19	15	0.3	300	2.5	4.6	5.0	5.5	6.1
5	0.12	15	1.0	500	2.5	6.4	6.9	7.4	8.0
6	0.19	15	1.0	500	2.5	6.9	7.4	8.1	8.8

Notes
1 A single 600 × 115 door 400 mm above ground level and a 168 mm diameter base, $K = 2.2$.
2 Material 275 N/sq mm. Ignoring shaft, bracket and deflection limitations.

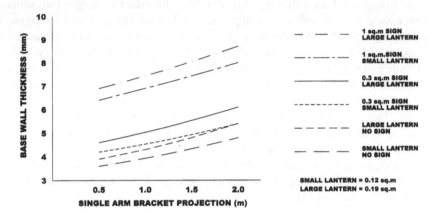

MINIMUM TUBE WALL THICKNESS IN DOOR AREA FOR STEEL GRADE S275 FOR A TYPICAL 10m COLUMN WITH BASE DIAMETER 168mm & k=2.2

Figure 3.43 Minimum tube wall thickness in the door area for steel grade S275 for a typical 10 m lighting column with base diameter 168 mm and $K = 2.2$.

fittings, and it will not generally be necessary to have a projection bracket so common for street lighting.

Poles will, therefore, frequently be designed in the same way as a post-top lighting column, and all of the clauses in 3.6.4 will apply.

3.6.6 Hinged lighting columns

3.6.6.1 General

Hinged lighting columns offer the same service as lighting columns in providing a support for the luminaire(s), and must meet all the requirements in Clause 3.6.4, but in addition they provide a means of lowering the luminaire(s) to an accessible height for ease of maintenance.

The hinging design and operation lies outside the requirements of BS EN 40, and other structural, mechanical and electrical considerations apply. Some of these are indicated in the next two sections for the two main types of hinged lighting column, but at any hinge safety of operation and appropriate protection for the operator is essential, as well as the protection to electrical cables flexing or passing the hinge point.

3.6.6.2 Mid-hinged lighting columns

There are a variety of designs of mid-hinged lighting columns, mostly of octagonal section although other sections are possible. The general concept is to provide a fixed part of the lighting column, a little over half the overall height installed vertically in the ground, and a moving part pivoting in the central portion; rather like a playground see-saw. The moving part will have the luminaire at one end, and a fixing point and possibly additional weights at the other known as the 'tail'. When in service the moving shaft will be vertical with the luminaire at the top, and the tail securely fixed to the fixed vertical lower part of the lighting column.

To lower the luminaire for maintenance, the tail will be released and restrained by a rope so that the slightly heavier end (where the luminaire is attached) drops under control to a position near ground level.

The rope is generally restrained manually as the moving shaft is equally balanced, but, dependent upon size and weight, other mechanical aids may be provided to lower and raise the moving shaft.

Safety is of paramount importance due to the inevitable possibility of trapped fingers or collisions with operators when moving parts are employed. Careful consideration has also to be given to the electrical supply cable and the need for restraint and flexibility at the hinge point.

3.6.6.3 Bottom-hinged lighting columns

Bottom-hinged lighting columns have a base flange or base section which is fixed in the ground or on a foundation in the normal way. To this plate or base is fixed the remainder of the lighting column by means of a hinge. The whole of the upper part of the lighting column can be hinged down for maintenance of the lighting.

As the lighting column is not substantially in a state of equilibrium like the mid-hinged lighting column, some additional restraint has to be provided to safely control the lowering and raising operation. This can be by means of internal springs or hydraulics, or by an external removable tool operated by hydraulics or wire ropes and winches.

Allowance has to be made in the design of the foundation for a bottom-hinged lighting column for the eccentric load imposed during lowering. This load may exceed the in-service wind load for which the foundation would be designed if it were a static lighting column.

As with the mid-hinged lighting column, safety during operation is of paramount importance.

3.6.7 High masts

3.6.7.1 General

High masts are generally of greater height than lighting columns or poles, most frequently 18 m minimum. High masts are regarded as a specialist area and there are no UK or European Standards. There are a variety of structural standards that are occasionally specified covering towers, chimneys or masts, but these are not specific to the 'soft' or 'flexible' dynamic design that is used for high masts. The main, if not the only, complete specification covering structural design, erection, operation and maintenance is the ILE Technical Report No. 7. There are two versions currently in use, the Second Edition based on gust-wind speed and the later 2000 Edition based on mean hourly wind speed, which will eventually replace the earlier version. There is some overlap in height terms, as these documents give guidance on design from 10 to 60 m height.

High masts can be fixed head where maintenance access has to be provided, or raising and lowering, either by hinging or by vertical raising and lowering of the luminaire carriage on wire ropes.

3.6.7.2   Fixed head masts

For fixed head masts, the luminaire array will be secured at the top of the mast on beams or in a frame, with or without one or more working platforms. Much larger configurations can be accommodated without weight or windage limitations of lowering equipment. This type is, therefore, widely used for sports stadia, where the luminaire array can be arranged on a vertical continuation of the mast or on an inclined section at the top. However, the angle of tilt should be limited as far as possible to avoid the inevitable deflection and curvature of the vertical part of the mast under this additional eccentric loading forward of the mast axis.

Access options are

- External by means of a mobile hydraulic access platform
- By means of internal or external ladders, fixed or removable rungs with appropriate safety devices
- By an access cage hoisted on wire ropes or self-climbing on the mast or on a fixed track.

3.6.7.3   Raising and lowering masts

Possibly the most common method to give access to the luminaire array on high masts is by means of a vertical raising and lowering device. The luminaires are arranged on a carriage suspended by wire ropes which pass over pulleys in the head frame at the top of the mast and are secured in the base compartment just above the base flange. They can be secured to the drum or drums of a permanent winch, allowing direct raising and lowering of the carriage by operating the winch, or to a termination point to which the wire ropes or chain of a removable winch can be attached to achieve raising and lowering operations.

In design, consideration has to be given to the flexibility of the power supply cables and to their ability to support their own weight over the height of the mast. The wire ropes should be of consistent construction with limited stretch or flattening characteristics.

The numbers of supporting ropes and winch drums will vary for different designs. Two or three rope suspension systems using one, two or three winch drums are available. However, tracked systems also exist with different driving and supporting systems.

With wire-rope systems there are also two main types of support in service, those with latches and those without. The 'in tension system' does not rely on latches to hold the luminaire carriage against the docking point at the head of the mast, as the wire ropes are constantly in tension throughout their life. Even if latches are fitted, they act as a safety device if one or more ropes should fail. Other latch designs require the ropes to be taken out of tension. Due consideration should be given to the life of the ropes and likelihood of kinks and also the maintenance of catches and the consequences of their failure to latch or to unlatch. The life of a rope is determined by the stress cycle throughout its life, and a rope in almost constant tension will last considerably longer than one that is frequently de-tensioned and re-tensioned to the same load.

The wire-rope construction and material is also very important dependent on the details of the actual system design. Galvanised rope will suffer from corrosion eventually, and there is the possibility of greater corrosion at points of contact with other materials. Stainless steel ropes are far less vulnerable to corrosion or interaction and are preferable. Flexible ropes should always be used and the construction of the strands may be critical as many rope constructions utilise compressible fibre cores where flattening of the rope under load over a pulley or in multiple layers on a winch drum may give rise to problems, as compared to 'solid' construction ropes, that is ropes using wire cores.

3.6.7.4 Hinged masts

These masts bring the luminaire array down to ground level for maintenance by means of a hinge at or near ground level. An external raising and lowering tool is generally required and can be either hydraulic or a system of wire ropes and winches using levering principles. In practice, due to the much greater loads for high masts than for lighting columns, the hydraulic tools are more practical.

For this type of mast, maintenance and aiming of floodlights is carried out with the luminaire array in a plane at right angles to the in-service position. Due allowance must also be made in the foundation design to avoid excess settlements in the direction of lowering.

3.6.7.5 Safety

Safety is of paramount importance for the access and operation of high masts due to the height involved. Training of personnel is also very important to ensure familiarity with the designer's operational and maintenance procedures. National Health and Safety requirements may vary and should be complied with, such as the UK's LOLER regulations, together with any particular requirements for wire ropes and hydraulics. However, any regulations should be applicable to high mast structures and their frequency of use. Requirements for equipment such as building lifts should not be automatically applied to high mast structures.

3.6.8 Catenaries

Catenary lighting can be a very efficient method of lighting major roads and indicating to the driver the direction of the carriageway. The poles and suspension system have very specific requirements that are not covered by any one particular standard. They are more common in some parts of Europe than the UK, particularly Belgium where slender designs are in use. In the UK the structures have tended to be of heavier construction due to the Highways Agency's deflection requirements.

Two main types are in use, 'fixed spans' and 'running spans'. With fixed spans the supporting catenaries are fixed to the poles at each end, whereas with running spans the cables pass over pulleys at the top of several intermediate poles to create a multi-span structure. The loading on intermediate poles is thereby reduced to result in lighter and more economic poles. Adjusting the catenary spans can be a time-consuming operation to ensure consistent sags and tensions as the lighting columns flex under each span tension change. Due consideration has to be given to impacts and the possible removal of a single pole, particularly for running span designs.

3.6.9 Corrosion protection

Cosmetic appearance is not the same as corrosion protection, and additional cosmetic painting may well be required in some locations at a much higher frequency than necessary for corrosion protection.

With the exception of concrete, aluminium, stainless steels and composite lighting columns, a good protection system is required to prevent corrosion and loss of section and structural strength. Any protection system must be appropriate to the location and to the degree of attack from the environment. In many non-coastal areas hot-dip galvanising to BS EN ISO 1461 will provide adequate protection even without additional painting for 25–100 years. In coastal and polluted areas an additional paint coating will prevent the loss of zinc and provide sufficient protection possibly for the life of many lighting columns and masts. Galvanising provides a very practical sacrificial protection, which is also economic for lighting columns, poles and masts, and provides reliable internal protection.

Many other systems are available but cannot always be applied to the internal surface. Without internal protection the life of the structure will be shortened by the rate of unseen internal corrosion.

3.6.10 Foundations

Lighting columns and poles are usually either planted (rooted) for direct burial or flange plated.

Rooted lighting columns have to be backfilled over virtually the whole length of the root either with good well-compacted soil or with a lean-mix concrete. Good guidance on the root surround requirements is given in the National Appendix to BS 5649: Part 2. This gives a formula for calculating the diameter of hole and hence the concrete surround necessary for a given lighting column and planting length based on three types of soils (good, average and poor).

Flanged lighting columns are connected to a reinforced concrete foundation by means of bolts. The foundation has to be designed for the forces created by the lighting column under the extreme wind conditions and the size will be very dependent upon the prevailing soil conditions. Additional loading cases may well have to be considered in design, particularly for bottom-hinged masts or lighting columns (see Sections 3.6.7.3 and 3.6.7.4).

Flange plates can be mounted on shims and then grouted or on levelling nuts when the gap under the flange can remain open or filled with a non-structural grout. The designer will advise on the appropriate installation for a particular design.

3.6.11 Inspection and maintenance

As most lighting columns, poles and high masts are in public areas, their structural integrity is very important to ensure that premature failure due to corrosion or damage does not give rise to personal injury.

An appropriate regime of inspection and testing should be defined, maintained and recorded to ensure that potential problems are identified in good time so that remedial action or replacement can be arranged.

For lighting columns, common problem identification and a suggested regime is given in ILE Technical Report No. 22, which is also appropriate for poles.

For high masts, ILE Technical Report No. 7 proposes a sequence of inspections, which may be modified for a particular design.

In all cases a separate inspection and painting regime may well be required to maintain the visual appearance of the structure.

3.7 Maintenance

3.7.1 Introduction

Lighting installations must be maintained to prevent excessive deterioration of light output. Lamp flux falls continuously through the life of a lamp until failure occurs. In luminaires light is lost by absorption due to dirt accumulating on the outside and inside of the visor and the optical system. In addition there are structural and aesthetic issues associated with the supports, and electrical safety to consider.

The designer should prepare a schedule to guide future maintenance. There are two categories of maintenance: reactive and preventative. One should not be used in preference to the other: a combination of both will give the best results.

3.7.2 Maintenance schedule

If an adequate maintenance schedule is not prepared there are likely to be delays in renewing faulty/damaged equipment, and additional costs to the maintaining body.

The maintenance schedule should include the following:

- Schedules of installed equipment together with plans cross-referenced to show locations
- Schematic diagrams showing cable sizes and earth fault loop impedance and voltage drop values if applicable
- Duct location plans
- A copy of the specification
- Manufacturers' technical data sheets and recommended maintenance details if applicable
- For more sophisticated equipment, any codes or programming details
- The recommended preventative maintenance periods
- Schedules of re-ordering codes and supplier contact details
- The electrical inspection and test certificates
- Structural test certificates if applicable.

3.7.3 Reactive maintenance

3.7.3.1 Lamps

For lamps a reactive maintenance policy is known as burn to extinction. The strengths of this policy are

- Possibly cheaper
- Lamp life is maximised

- Reduced environmental impact in respect of glass and mercury disposal
- Automatically takes account of improved lamp technology.

However, there are attendant disadvantages:

- Increased loss of light output through life
- Increased lamp failures
- Increased control gear failures
- Increased administration
- Design standards are not achieved (i.e. inefficient use of capital funds)
- Reduced benefit to the user[1] (both from failed lamps and lower output)
- Reduced light output for the same energy cost (for non-SOX lamps).

The use of a burn-to-extinction policy only is not recommended, as it tends to lead to an uncared-for-looking installation with widespread dark patches, giving the impression the owner does not care about the appearance.

3.7.3.2 Luminaires

Luminaire failure is usually dealt with on a reactive basis.

3.7.3.3 Control gear

Control gear failure, other than for capacitors, is usually dealt with on a reactive basis.

3.7.3.4 Visual inspections

A visual inspection should be carried out at each maintenance visit. The inspection should identify

- Damage to electrical terminations
- Structural deterioration of the support, including accident damage
- Damage to the luminaire
- Integrity of any cosmetic treatments
- Presence of unwanted attachments to the lighting columns.

3.7.3.5 Painting of supports

If painting of supports is left until it is reactive, that is the loss of paint adhesion or the formation of corrosion, the appearance of the installation becomes unsatisfactory. It is recommended that painting is carried out on a preventative basis.

3.7.4 Preventative maintenance

3.7.4.1 Lamps

Preventative maintenance of lamps is generally carried out by group replacing all the lamps in a system at set intervals to minimise outages and premature failures whilst maximising the useful life of the lamps. The strengths of a group lamp replacement policy are as follows:

- Consistent lighting quality
- Reduced lamp failures
- Easier administration and planning
- Reduced user defect notifications
- Benefit to the user maximised
- Lamp renewal can be integrated with other planned maintenance.

However, there are attendant disadvantages:

- May be a more expensive policy
- Lamps replaced before failure
- Reliance on manufacturers' data[2]
- Increased environmental impact in respect of glass and mercury disposal.

The benefits of a preventative maintenance strategy for lamps and luminaires are demonstrated in Figure 3.44.

For best results a combination of preventative and reactive maintenance is recommended, i.e. lamps and luminaires should be subject to planned maintenance at pre-determined periods, with any failed lamps in the intervening period replaced.

3.7.4.2 Luminaires

For luminaires having an IP rating less than IP66, dust and/or dirt will accumulate on the inside and outside of the luminaire. Reflective elements and the internal and external surfaces of the visor should be cleaned using non scratch products. For luminaires having an IP rating of IP66 or greater, dust and/or dirt will accumulate on the external surfaces of the visor. This should be cleaned using non scratch products.

Where birds use luminaires as a perch, corrosive, unhealthy and unsightly deposits are likely. Operative wearing suitable protective clothing should remove these deposits. Where such deposits are excessive consideration should be given to the attachment of anti-bird devices to the top of the luminaire.

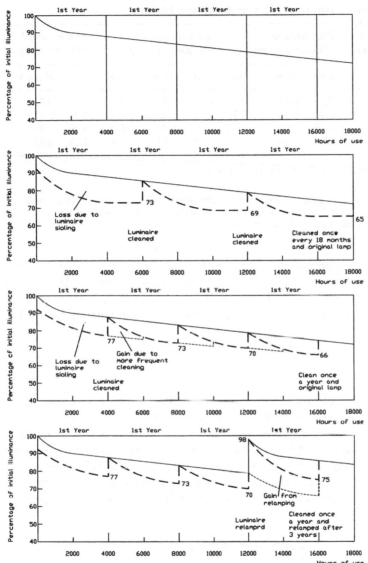

Figure 3.44 Changes in illumination with different maintenance strategies.

Table 3.17 Luminaire maintenance factors

Proposed cleaning interval (months)	Ingress protection number of lamp housing								
	IP2X minimum (Pollution category)			IP5X minimum (Pollution category)			IP6X minimum (Pollution category)		
	High	Medium	Low	High	Medium	Low	High	Medium	Low
12	0.53	0.62	0.82	0.89	0.90	0.92	0.91	0.92	0.93
18	0.48	0.58	0.80	0.87	0.88	0.91	0.90	0.91	0.92
24	0.45	0.56	0.79	0.84	0.86	0.90	0.88	0.89	0.91
36	0.42	0.53	0.78	0.76	0.82	0.88	0.83	0.87	0.90

Source: Reproduced by kind permission of BSI.

Notes
High pollution occurs in the centre of large urban areas and in heavy industrial areas.
Medium pollution occurs in semi-urban, residential and light industrial areas.
Low pollution occurs in rural areas.

The luminaire maintenance factor determined from Table 3.17 for the project should be multiplied by the lamp luminous flux maintenance factor for the type and rating of the lamp to be used at the proposed group lamp replacement frequency to obtain the overall maintenance factor for the lighting system.

3.7.4.3 Control gear

Capacitor failure leads to the inefficient use of energy and with discharge lamps will usually mean the power factor is outside the limits required by the electricity supply company, who may apply penalty charges. Consideration should be given to the cost-effectiveness of group replacement of capacitors. The frequency of group replacement of capacitors should take account of the anticipated life of the capacitor and the frequency of other maintenance operations. The recommended maximum life of capacitors is 6 years.

3.7.4.4 Electrical Inspection and testing

Electrical inspection and testing of electrical apparatus in lighting systems should be carried out in accordance with BS 7671. The recommended maximum period between periodic inspections and testing for Highway Equipment is 6 years, to align with other operations (such as group lamp replacement) and to minimise inconvenience to the road users. Further information about the recommended frequencies between periodic inspections and testing may be found in Guidance Notes to BS 7671 published by the IEE and in the ILE Code of Practice for Electrical Safety in Highway Electrical Operations.

3.7.4.5 Visual inspection

A visual inspection of the equipment should be carried out at each maintenance visit. The inspection should identify

- Damage to the electrical systems
- Structural deterioration of the supports, including accident damage
- Damage to the luminaire
- Integrity of any cosmetic and/or protective treatments
- Presence of any unauthorised attachments.

3.7.4.6 Painting of supports

The frequency of support painting should only be extended beyond 6 years where applications specifically formulated for longer life have been applied.

3.7.4.7 Structural

All supports will deteriorate and will require inspection, maintenance and eventually renewal. The life of the support, the frequency of maintenance and the type of inspection vary according to the

- location
- use
- design and
- material

of the support in question. In recent years, commercial pressure and greater sophistication in design have resulted in supports which are more closely matched to their design-loading requirements, without additional strength that may have been present in earlier designs. The danger of this trend is that undetected corrosion can cause premature failure. Significant defects for different lighting column types are tabulated in Table 3.18.

In the UK the most common supports are steel lighting columns. For steel lighting columns the amount of corrosion is dependent upon

- The effectiveness of the protective treatment provided to both the internal and external surfaces of the lighting column, and the root
- Sodium chloride in de-icing salts and/or sea water
- Dog urine
- Atmospheric pollution and acid rain

Table 3.18 Significant defects for different lighting column types

Significant defects	Types of lighting column					
	Tubular steel	Sheet steel	Aluminium	Concrete	Composite	Cast iron
Corrosion up to 300 mm above ground level			*			
Corrosion from the bottom of the door to just below ground level	*	*				*
Corrosion of the root of planted columns	*	*	*			*
Corrosion at the door opening	*	*	*			*
Cracking at the door opening	*	*	*		*	*
Internal corrosion of overlap joints	*	*				
Cracking below ground						*
Cracking at welded joints	*	*	*			
Failure of adhesives			*		*	
Corrosion of pre-stressing wires of pre-stressed columns				*		
Corrosion of the reinforcement and cracking of the concrete at the base and around the door opening				*		
Corrosion of reinforcement and cracking of concrete at or on the bracket/shaft joint				*		
Impact damage and vandalism	*	*	*	*	*	*
De-lamination					*	
Effects of UV					*	

- Chemicals such as weedkillers
- Ground-water retention
- Aggressive ground conditions, for example sulphates, sulphides and chlorides
- Stress caused by wind exposure
- Additional stress caused by attachments.

Testing methods and further information are given in the ILE Technical Report No. 22.

Chapter 4

Techniques for particular applications

4.1 Introduction

The principles of lighting apply equally to the outdoor and indoor environment. However, the outdoor environment is subject to influences that either do not exist or have very much lower significance indoors. If the impact of these issues is underestimated the effect can be a combination of the production of unacceptable glare, intrusive light, excessive capital and maintenance costs and significant non-performance of the lighting installation.

The principal influences to be considered are outlined in Table 4.1. Guidance on the influences is given in Section A.3.1.

The following sections give guidance on best practice for a variety of outdoor locations, together with recommended lighting levels.

4.2 Town and city centres

4.2.1 Introduction

Over the last 50 years, the engineering principles of lighting for traffic have dominated the design of public lighting. Even the lighting of commercial streets in town and city centres, which are either fully pedestrianised, or are primarily used by large numbers of pedestrians at night, are lit using these principles. As a result the lit scene in many urban centres is bland and uninteresting, with poor modelling of architectural features, harsh sometimes monochromatic colour effects and a high degree of glare. However, BS 5489-1: 2003 states

> In urban and amenity areas the efficient lighting of the road surface for traffic movement is not the only or even the main consideration. A balance with many other aspects has to be achieved. Urban centres serve many users, each with differing and sometimes conflicting needs. A balance with many other aspects therefore has to be achieved. A master plan should be drawn up which contains the relevant objectives in order of their perceived importance and emphasis.

Table 4.1 Influences on lighting of the outdoor environment

Influence	Issues to consider
Weather	Water ingress Drainage of water Operational temperature Ease of maintenance in temperature extremes Ability to withstand wind pressure Effect of mist, fog, blizzards, etc.
Corrosion resistance	Protective systems Electropotential difference Acidity/alkalinity
Vandalism	In most areas, open access for vandals
Equipment design	Surface temperature
Glare	Design of luminaries Aiming of luminaries High contrast
Light pollution	Design of luminaries Aiming of luminaries Most upward light is incident on other planets Intrusive light
Aesthetics and visual intrusion	Daytime appearance Night-time appearance
Electrical design	Voltage drop Cable sizes Protection
Vehicles	Damage to lighting apparatus
Reflection from surfaces	No ceilings Limited perimeter walls, fences enclosures, etc. Low surface reflectance
Lighting-system user	Multiplicity of viewing angles Mobility of user Speed of user
Lighting-system observer	Multiplicity of viewing angles from outside lit area Mobility of observer Speed of observer

Insufficient cognisance is taken of this.

If modern town and city-centre areas are to be made safer and more secure and to sustain a thriving 'evening economy', the visual needs of pedestrians have to be addressed, through the adoption of a range of modern lighting techniques and technologies.

4.2.2 The visual needs of pedestrians

Lighting requirements for pedestrians are radically different to those of the motorist. Notwithstanding that pedestrians have an easier visual task, to date there has been little research done on their visual requirements compared to those of the vehicle drivers. The following guidelines for designing pedestrian-oriented, pedestrian-friendly lighting are therefore based on current research into 'visual interest' and 'visual comfort' in interior-lighting schemes and practical experience of the way that people use urban centres at night.

4.2.2.1 Vertical illumination

The Bartlett School of Architecture has undertaken research into subjective preferences for different kinds of lighting. Loe, Mansfield and Rowlands[1] established the crucial role of the 40° viewing zone as the primary zone of 'visual interest'. This is defined as a cone 20° above and below a horizontal line extended out from the eye (Figure 4.1).

For a pedestrian in a typical urban environment, this viewing zone will include the principal vertical surfaces around them (walls, trees, monuments

Figure 4.1 Primary zone of 'visual interest'.

and other people). Therefore, care should be taken to ensure these vertical surfaces are illuminated.

4.2.2.2 Visual interest and visual brightness

The difference between an object or a surface which is about twice as bright as the adjacent area is just noticeable, whereas a luminance ratio of 5:1 appears significantly different and a ratio of more than 10:1 in apparent brightness is emphatic and the lighting on such a feature emphasised to this extent might be called 'dramatic'. Loe *et al.* established that lit scenes perceived as visually interesting and having visual brightness had maximum/minimum luminance ratios of 13:1 or greater. Whilst we may wish to introduce visual interest to our town and city centres, care must be taken to ensure that adequate illumination is still provided so people are visible and their intention can be assessed. Large variations in lighting levels may cause problems for CCTV cameras which may not be able to adjust to the extremes whilst panning or zooming.

4.2.2.3 Glare

The problem of glare to pedestrians is an issue that has not been researched. However, perceived glare, both conscious and unconscious, could be discouraging people from using city-centre streets at night, which may have significant commercial and social repercussions.

4.2.2.4 Colour rendering

If one considers the activities of pedestrians involved in the 'evening economy' – street markets, outdoor eating and drinking, clothes shopping, street entertainment, meeting friends and 'people watching' – one can see that there should be a high premium on lighting with good colour rendering capabilities ($R_a 80$ or above). People need to be able to see such things as skin tones, planting, fabrics, food and drink in a way that approximates to their 'natural' colours.

4.2.3 Traffic lighting versus pedestrian lighting

The uniform visual effects of the road lighting design approach can be seen in Figures 4.2 and 4.3, one from a local public-lighting scheme in the UK and the other from a prestigious square in the centre of Paris. Despite their differences in budget and location, they exhibit the same features, high horizontal illuminance, low vertical illuminance with poor modelling of the surrounding buildings, harsh orange-tinted lighting with poor colour rendering and high uniformity creating a bland, uninteresting visual scene.

Figure 4.2 Local pedestrian area lighting (see *Colour Plate 4*).

Figure 4.3 Place Vendome, Paris (see *Colour Plate 5*).

The night-time visual amenity of such areas could be greatly enhanced by the lighting of the surrounding buildings, trees and monuments.

Unfortunately, such lighting is often outside the remit of the authority responsible for the highway and its lighting. Lighting practitioners are recommended to encourage building owners etc. to light their properties to complement and enhance the town- or city-centre night scene. Such lighting should be provided in a coordinated manner to complement and not distract from the overall visual scene. Early consultation is recommended to ensure the compatibility and suitability of individual items and the overall scheme.

4.2.4 The role of anti-pollution measures

The last few years have seen an increase in resistance against light pollution in a number of countries, most notably the USA, the UK and Italy. Direct and reflected light illuminates dust and water particles in the atmosphere and leads to extensive 'sky glow' over many of our urban centres, blocking out our view of the stars. For more information, refer to Section 2.3.

Various measures have been adopted to reduce the impact of light pollution. The most notable of these being the adoption of 'full cut-off' luminaires, which project little or no light above the horizontal. However, this can further exacerbate the tendency to over-light horizontal surfaces, while neglecting the all-important vertical surfaces, such as walls, planting and monuments.

For all their failings, one of the side-effects of old-style luminaires was that at least they gave some back-spill onto the upper facades of buildings. No matter how unplanned or accidental, this wash of light did help to define the surrounding architecture and give it visible form. Alternatively, depending on the height of the light columns, we may see a shadow-line part way up the façade. The best way to ensure high night-time amenity and visual interest is to provide individual lighting schemes specifically designed to complement and coordinate with an overall composite concept. In this way each lighting scheme can be optimised for its purpose whilst still forming part of the overall lit scene without conflicting with other areas.

Of course, no one is advocating a return to the use of 360° distribution globe fittings, so popular in the 1960s and 1970s, but the adoption of tighter luminaire controls places greater onus on lighting practitioners to mitigate their effect. Ingrained lighting design policies need to be re-thought and property owners need to be encouraged to light their buildings in a sensitive and coordinated manner to help re-balance the night-time scene and to give our town and city centres a fresh emphasis.

This can be done in a number of ways, for example

- Architectural lighting units mounted on the back of lighting columns or integrated into luminaires
- Adjustable purpose-designed louvre systems incorporated in to the luminaire design, so that 'back-spill' onto nearby walls can be carefully designed out on site
- Dedicated façade lighting by building owners, encouraged by local authorities.

4.2.5 New lighting in practice

David Loe and Peter Tregenza in their book *The Design of Lighting*[2] state

> Clear visibility is necessary for good security. A well lit area receives less criminal activity than one that is dark and gloomy. Perhaps even more important is that people feel secure in an environment that is reasonably bright. To achieve this there must be continuity of illumination – no dark zones along a pedestrian route if personal attack is a possible or perceived hazard, no unlit areas around the perimeter of a building where burglary might occur.

Based on the above principles, it is possible to summarise a number of lighting design techniques and technologies, which can be combined to create the kind of lighting scene that will provide visual comfort and interest for pedestrians.

These include

- The replacement of conventional, sodium-based luminaries with modern 'white' light sources (ceramic metal halide or fluorescent) – there is a substantial body of evidence[3,4] emerging that at (mesopic) night-time light levels, 'white' light can be far more efficient in visual terms than limited-spectrum light sources
- The lowering of mounting heights to a more human-based scale commensurate with the surrounding architecture
- The rejection of high overall uniformity of both luminaire lay-outs and resulting illuminance, in favour of lighting arrangements that are more varied and less predictable, but retain safe pedestrian routes without dark spots
- A greater use of indirect lighting techniques, to create a softer, glare-free, more visually comfortable environment
- More extensive lighting of building façades, trees and planting, to increase the emphasis on vertical surfaces

- The careful, selective use of coloured lighting, to add a degree of visual interest and excitement to our urban environments
- Ultimately, with the latest lighting technologies (moving head luminaires, movement sensors, the dimming of discharge lighting and so on) we could even see 'interactive' lighting systems that respond to the presence and movement of pedestrians, to create a more dynamic, less static night-time environment.

4.2.6 Conclusions

For commercial and social reasons, many UK town and city centres are working to make themselves more pedestrian-friendly and less dominated by traffic. A new approach to lighting, derived from an analysis of the visual needs of pedestrians, could play a contributory role in this long-overdue process. Careful use of new lighting techniques and technologies could contribute greatly to making people feel safer and more secure, encouraging them to come out at night, stay in the streets longer, and shop and dine later into the evening. All this will help to create conditions in which crime and the fear of crime is reduced.

The following is a checklist of functional requirements for town and city centre which is recommended for consideration when lighting these areas:

- Safety
- Security
- Guidance and orientation
- Glare, distraction and light pollution
- Luminance
- Colour
- Installation
- Energy conservation.

4.2.7 Recommendations

For roads within city and town centres which primarily carry traffic, refer to Table 4.29.

Table 4.2 gives recommended lighting classes for pedestrian areas and mixed vehicle and pedestrian areas in city and town centres. The choice of lighting class for a specific city or town centre road type may be varied up or down from the classes proposed taking account of

- Vehicular traffic use
- Pedestrian and cyclist use
- On-street parking.

Table 4.2 Lighting classes for pedestrian and mixed vehicular and pedestrian areas in city and town centres

Pedestrian traffic flow	Normal		High	
Environmental zone	E3	E4	E3	E4
Pedestrian only traffic	CE3	CE2	CE2	CEI
Mixed vehicle and pedestrian with separate footways	CE2	CEI	CEI	CEI
Mixed vehicle and pedestrian on same surface	CE2	CEI	CEI	CEI

Source: Reproduced by permission of BSI.

In addition other amenities such as shops, public houses, etc. and the level of crime may influence the choice of lighting class.

4.3 Lighting and closed circuit television (CCTV)

4.3.1 Introduction

As most lighting practitioners are not closely aware of what CCTV encapsulates and what technical constraints are applicable it is worth reviewing such information. This section briefly touches on the uses of CCTV and the basic system set-up, before dealing with lighting issues.

4.3.2 CCTV applications

Close Circuit Television technology is perceived as a valuable loss prevention and safety and security management tool. Police and Local Authorities use CCTV as a deterrent to criminal activity and to monitor and record anti-social behaviour and general criminal activity. CCTV is used to identify visitors and employees, monitor hazardous work areas, thwart theft and ensure the security of premises and parking facilities.

4.3.2.1 Security applications

Security applications are to

- Observe and record theft or violence by overtly monitoring open public space and street scenes, retail floor space, office buildings, building perimeters, warehouses, loading docks, and car parking
- Observe and record shoplifting activities

- 'Walk a beat' by programming a moving camera to pan, tilt, and zoom within a defined pattern
- Perform covert surveillance (where legally applicable)
- Integrate with access control systems to provide video of persons entering and leaving premises
- Complement asset tracking systems to provide video when a tagged asset leaves the premises.

4.3.2.2 Safety applications

Safety applications are to

- Allow operators to see into areas where the environment is hazardous to life or health (i.e. hazardous materials, chemical toxins, etc.)
- Monitor potential accident areas
- Monitor common areas, or high-risk areas to ensure safety of an educational institution's students and faculty
- Help reduce the severity of some incidents by the timely dispatch of the emergency services.

4.3.3 CCTV systems

A CCTV system consists of the following elements:

- Environment
- The cameras
- Camera spectral response
- The lens
- Video transmissions methods
- The monitor
- Peripheral equipment.

Although it is not appropriate here to consider the detail of the CCTV system it is worth developing a few of the above terms that are affected by lighting.

4.3.3.1 The environment

The environment often contains different colours, surfaces and materials that reflect varying levels of light. To select the appropriate CCTV equipment, it is necessary to determine the minimum lighting level (day or night) that will arrive from the scene to the camera lens. The 'available' light will affect everything from picture clarity to focus.

During the hours of darkness the area can be illuminated by artificial light sources such as incandescent, gas discharge, IR lamps, or other artificial lights. It is said that an axiom in CCTV security applications is: The better the light, the better the picture.

4.3.3.2 The cameras

All cameras need light that must be of the appropriate colour and levels to give good pictures.

- Black and white cameras will work with most forms of lighting and also with IR lighting, but be aware that IR lamps have a practical range of about 80 m. Beyond that, extra IR lamps or other lights would be needed.
- Colour cameras need more light than black and white cameras, and the light must be white rather than the common orange low pressure sodium colour, to maintain correct colours in the picture. Most colour cameras will not respond to IR lighting.

Camera specifications will often claim to work 'down to 0.5 lux'. However, this does not take account of the lens used or the subject area material, e.g. grass is more reflective than tarmac.

As a general rule the horizontal illumination of a minimum of 5 lux for black and white cameras and 15 lux or better for good colour pictures is required.

There are some colour cameras that will work with IR lamps, by switching from colour to black and white mode automatically at dusk, and back to colour again at dawn. These can be useful for car parks or general perimeter areas where a high level of local white light is not currently practicable or economical.

4.3.3.3 Camera spectral response

Similar to the human eye, cameras have different responses to the spectrum of light. This is usually shown in diagrammatic form and is known as the relative spectral sensitivity. Figure 4.4 shows the relative response to each part of the spectrum for two cameras. It can be seen that camera A covers the visible part of the spectrum very effectively whereas camera B is sensitive far into the IR. It could be thought that camera B would suit all requirements because of the wide range of wavelengths covered, but not so. A further point is that using an IR sensitive camera in daylight can produce different ranges of grey tones because they see a higher content of IR than the eye. Also the IR sensitive camera can cause the automatic

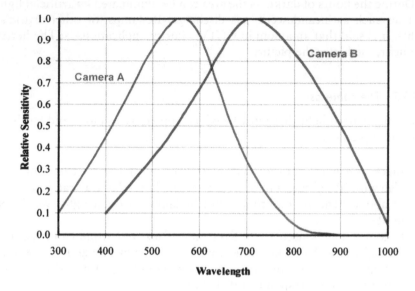

Figure 4.4 Relative spectral response of two cameras.

iris to close due to the amount of IR light instead of visible light. This is particularly noticeable if there is foliage in the scene. Chlorophyll reflects at about 715 nm and often appears bright white instead of a shade of grey. In practice, the sensors sensitive to extended IR light are more sensitive than photopic sensors.

Incidentally, all camera sensors are monochrome; colour is obtained by inserting red, green and blue filters in front. This is why colour cameras have less resolution than monochrome cameras. Also due to the filters, colour cameras are not sensitive to IR light. Therefore all the discussion on camera sensitivity and suitability for IR illumination is confined to monochrome cameras. Dual mode cameras now becoming available potentially offer the best of both worlds. This, however, is evidence of the fact that colour cameras cannot be as sensitive as monochrome.

4.3.3.4 Lenses

The first component that light from a scene has to pass through is the lens. If this is wrong then everything after gets progressively worse. Apart from selecting the focal length for a particular scene, there is generally very little thought given to this prime element in the chain to get a picture from a scene to a monitor. There are several factors in lens selection that will affect the effectiveness of a system.

The four main factors are

- Correct focal length or range for the location
- 'f' number
- Resolution
- Spectral transmission.

Often they will determine whether anything at all will be seen.

4.3.4   Recommendations

4.3.4.1   Light and illumination

Only natural light provides near even illumination, as clouds and shadows affect it. The night-time scene suffers from the fact that as the distance from the light source increases so the illuminance reduces due to the inverse square law. The other effect of this is that the wide range of light levels can cause problems with automatic iris lenses. Unless set up correctly, the foreground light will cause the iris to close and detail in the background will be lost due to the reduction in light passing through the lens. The reverse is if the iris is set to the distant light level in which case there will be a lot of flare in the foreground.

4.3.4.2   Light levels

The above clearly indicates that to achieve a quality picture from a CCTV system during the hours of darkness a good-quality lighting scheme is required. That is one that achieves the correct minimum level of illumination with a high level of uniformity and a low threshold increment. In essence, where the CCTV is in the outdoor environment, in particular the street scene, it needs the same lighting conditions as the road user requires. Therefore the application of the relevant BS 5489 and corresponding EN standards will achieve an adequate level of illumination for most CCTV systems within the street scene. Recommended lighting levels for both black and white, and colour CCTV cameras are tabulated in Table 4.3.

Table 4.3 Recommended lighting levels for CCTV cameras

Cameras	Maintained minimum point illuminance (in lux) not less than	Overall uniformity not less than	Threshold increment not less than (%)
Black and white	5	0.4	10
Colour	15	0.4	10

Table 4.4 Light sources for CCTV use

Lamp type	Colour rendering index (R_a)	Colour appearance	CCTV use
Low pressure sodium	0	Orange	Not recommended. Colours not recognisable
High pressure sodium	20	Golden yellow	Fair to moderate. Use for monitoring purposes only
High pressure mercury	40–50	Bluish white	Moderate. Use for monitoring purposes only
Metal halide	65–90	Crisp white	R_a < 80 Good. Use for monitoring purposes only R_a > 80 Excellent. Use where recognition and prosecution evidence required
Fluorescent and compact fluorescent	80–90	White – in varying hues	Excellent. Use where recognition and prosecution evidence required
Tungsten halogen	100	Crisp white	Not applicable due to poor energy efficiency

4.3.4.3 Colour

There are two indicators that provide a guide to the spectral distribution of the light emanating from a source. These are the colour rendering index and the colour appearance. For ease of reference a listing of lamps in common use in the exterior environment for use with CCTV with their typical colour rendering values and colour appearance are shown in Table 4.4. For colour temperature information see Section 3.2.

For security use see Section 4.14.

4.3.4.4 Equipment siting

In the exterior environment cameras are usually housed in weatherproof fittings and mounted out of reach of the vandal. To obtain a good photograph for facial recognition, cameras are usually mounted between 5 and 8 m above the ground. The cameras then 'look down' a street recording the subject. However, street lights mounted between 5 and 8 m height can cause major problems to CCTV systems as a bright source in its direct line of sight can partially obscure the image. Therefore it is best to avoid this situation and it is recommended to have the light source above the field of view. Therefore if the camera is mounted at 5 or 6 m then the lighting system should be mounted at 8 or 10 m. This should relieve the camera system of most potential glare problems and provide an enhanced lighting system in the street.

4.3.4.5   Glare

To minimise the glare entering the lens of the cameras it is recommended to use luminaries with a reflector optic and a low-profile glass that has a high level of control of upward/horizontal light.

Where luminaries with a deep bowl are used to distribute the light from the lamp these inevitably spread the light more evenly in the environment. This may well be what is required in a shopping centre to add a little 'sparkle' to the night-time scene. However, this can reduce the effectiveness of the CCTV system.

It may not always be possible to neither site the luminaries above the cameras nor utilise low-profile lanterns in the street; here it will be necessary to consider the siting requirements of both systems. Both sets of equipment often need to use the same location to optimally site their equipment. For example, at a T-junction in a small shopping area a camera will view down the main street and the adjacent side street. However, the street lighting codes of practice recommends a luminaire in the same locality. On-site practicalities need to be assessed and a pragmatic solution agreed.

4.3.4.6   Lighting equipment

It is worth recognising the pressure to minimise street furniture is at conflict with the demands of society to treat our high streets in a similar manner to indoor shopping malls. To this end lighting columns are often an ideal platform to house much more than a simple luminaire. When requested column manufacturers have designed and manufactured lighting columns to carry traffic signs and hanging flower baskets in addition to the luminaire. Where lighting columns are required to carry additional loads such as festive decorations, pennants and banners, telecommunications systems, data transmission and CCTV cameras full details of the size, weight and position of the attachment on the lighting column should be provided to the lighting column manufacturer to allow a suitable lighting column to be designed and manufactured.

However, to achieve a CCTV picture that will give sufficient resolution a firm platform is required. Therefore a much stiffer lighting column than would otherwise be the case is needed to provide the support to the camera, and the higher the camera the more rigorous the demands on the lighting column. A number of years ago, the Home Office stipulated that cameras shall have only 5° deflections in 60 mph winds to achieve a high quality of picture. This will require a lighting column specifically designed to carry the cameras that can also be used as a lighting column.

Due to the inherent danger from cross-phase electrical supplies all electrical apparatus inside or mounted on a lighting column should be fed from the same electricity supply and phase via a main double-pole isolator. The individual electrical feeds for the lighting and the camera system should be separately fused and isolatable above the main double-pole isolator to reduce the risk of accidental interruption. Where a separate control pillar is sited adjacent to the lighting column to house the essential electronic hardware for the camera system, care should be taken to ensure it is fed from the same electrical phase as the lighting column, and consideration should be given to it being fed via the lighting column to allow the supply to be isolated for maintenance purposes.

4.4   Transport interchanges

4.4.1   General

Transport interchanges vary considerably in their layout, building type and facilities, and lighting designers should take account of the location, usage and surrounding environments of these installations. Common to all of them is their proximity to the highway and thus due account must be taken of the particular roadway lighting requirements laid down in BS 5489. Designs for roadway lighting within transport interchanges should be based on illuminance criteria in accordance with EN 13201 – Part 2: Performance requirements. The values to be achieved should be chosen from the CE series of lighting classes depending upon the type of roadway and its level of use.

Users of these facilities will vary in their familiarity with the environment and their mobility and visual abilities, but common to all groups is the transitory nature of interchanges. For this reason it is important that the lighting, both by day and after dark, creates visual adaptation paths and assists orientation. The peak luminances of both daylight and artificial light sources should be controlled to avoid disability glare and to minimise veiling luminances on information displays. Many of the tasks within transport interchanges take place in the vertical plane, e.g. reading signs and timetables. For this reason, and to provide an enhanced feeling of security, vertical illuminance values should be given careful consideration.

As with all lighting designs, the consideration of surface finishes and reflectance values will be a major factor in determining the required illuminance. Luminance contrasts between adjacent surfaces should not exceed 100:1 at any point in the far field of view and be limited to 30:1 for close viewing distances ($\leq$2.0 m). The rate of change of luminance should be progressive, not abrupt. Where luminance contrast is used to highlight routes and specific areas (for example exits), a visual step change of luminance in the range 2:1 to 5:1 is required. It should be borne in mind that as adjacent areas may have significantly differing reflectances, some localised treatment of the lighting may be required to achieve the desired effect.

A good alternative to the use of luminance contrast is the introduction of a colour variation, which in its simplest form requires only a change of lamp type. More extreme changes using saturated colour can be disturbing, and use more energy, so care should be exercised when using this method. The colour and texture of surface finishes, suitably rendered using light, should also be used as a means of controlling the lit appearance of the installation.

The design of lighting installations that are adjacent to the highway should not create lighting which is distracting to drivers or which cause glare. Where bus stations, interchanges and light rail stops are fully covered over, lighting should not be at such a high level at night as to create visual adaptation difficulties for drivers; alternatively the lighting should be graduated over a distance to allow adaptation of the eye before entering the normal road lighting system. There should be adequate lighting for safe movement for both traffic inside and outside, and luminance contrasts should take account of movements between these areas.

4.4.2 Daylight

All transport interchanges should maximise the use of daylight. Where large covered areas employ glazed canopies or roofs, a passive method of glare control should be employed to ensure that vehicle and pedestrian movements are not impaired by sunlight penetration.

The areas within the scope of this section vary too widely in type to specify particular daylight factor values, and it may be impossible to utilise daylight effectively in some instances. Where daylight factors exceed 1per cent, some form of daylight-linked control of the electric lighting should be employed. Where daylight factors are below this value, some artificial lighting will be required at all times. For further details please see the Society of Light and Lighting (SLL) Lighting Guide No. 10 – Daylighting and Window 1999.

Where large artificially lit areas are open, or have glazed roofs, the effect of the artificial light on the night sky (light pollution) shall be taken into consideration at the design stage.

4.4.3 Colour

The colour rendering of signs, maps and displays is of great importance to their effectiveness, and light sources of R_a 65 or better should be used to enable these to be seen clearly. In large areas, energy considerations may dictate the use of light sources with lower colour rendering indices, but these should not be below R_a 25.

Research has shown that people in transport environments prefer warmer colour temperature light sources. For this reason, light source CCT should be 3500 K or lower.

4.4.4 Information displays, advertisements and CCTV

Many displays utilise screens of some form that have relatively low self-luminance, of the order of 100 cd m^{-2}, and visibility of these will be difficult if their surrounds are relatively bright or if they are subject to high levels of incident light. Fixed illuminated displays should be sited such that the effects of incident daylight and artificial light do not reduce their effectiveness.

Hoods or cowls around display screens are rarely effective in controlling contrast problems. A wide dark border offers a means of keeping screen contrast high relative to the immediate surroundings. Displays should be orientated away from direct sunlight or large areas of unobstructed sky. Unshielded artificial light sources should not be placed where they will appear in the field of view at normal screen viewing distances.

Poster and timetable panels should utilise localised lighting. A vertical illuminance of 100 lux minimum on these should be achieved at all times.

Illuminated advertising material should not conflict with the main lighting. The lighting of advertising material should be in accordance with ILE Technical Report No. 5.

The lighting requirements of CCTV equipment should be taken into account. Illuminance values of 10 lux (vertical) are the minimum requirement where CCTV is in use.

4.4.5 Controls

Controls should be provided for all areas in order to minimise energy consumption and light pollution. All buildings and luminaires fixed to them require controls to meet the requirements of The Building Regulations, Part L.

Suitable controls include manual switching (with or without group-control contactors), photoelectric cells, presence detection (for areas of low frequency occupation) and scene-set dimming (in conjunction with fluorescent lamps).

4.4.6 Emergency lighting

Where required by The Building Regulations or Fire Precautions Act, emergency lighting should be provided in accordance with BS 5266 Parts 1 and 7.

4.4.7 Heritage buildings

Many older structures, particularly adjacent to railway premises, have historic or architectural significance. Some are listed by English Heritage, Historic Scotland or the Local Authority, and designers should take care to ensure that lighting schemes compliment such buildings and meet with the approval of the relevant conservation officer.

Lighting hardware and its associated electrical supplies should be located discreetly and its installation should not permanently alter the fabric of the building.

4.4.8 Characteristics of different interchanges

4.4.8.1 Light rail systems

For the most part, light rail systems will run adjacent to highways lit to the requirements of BS 5489-1, and the intermediate stops should be treated as bus stops. More major stops will require additional lighting for safety and to aid orientation.

The lighting should not create disability glare for the rail drivers nor create adaptation difficulties for users. The stops themselves should be lit in a cheery and pleasant manner that creates a feeling of security. Light sources of good colour rendering (at least R_a 60) should be used, and most users prefer a warm colour appearance, of 3500 K or lower.

Lighting at light rail stops should not impair the visibility of the rail signals and the lighting of roadways adjacent to segregated light rail tracks should meet the requirements of Section 12 of BS 5489-1.

4.4.8.2 Bus and coach stations

This section relates to the needs of bus stands that are off-highway installations. Equipment should be selected based on its suitability for the environment in which it will be used.

The passenger circulation areas, entrances, exits, amenity and waiting rooms should be lit to 250 lux at floor level with a diversity not exceeding 0.16. Discomfort glare in waiting rooms and staff accommodation should be limited. In waiting areas immediately adjacent to bus parking/loading bays, care should be taken that the lighting of the waiting area does not adversely affect driver's vision.

The lighting in the area of ticket counters and information points should achieve an illuminance of 500 lux at the desk top.

4.4.8.3   Road/rail interchanges

These should be lit to the same standard as bus and coach stations in general. An important consideration will be the boundary between the individual parts of the interchange, as separate lighting specifications may be involved. The lighting scheme should guide people safely between the various areas without stark contrast.

Drop-off points and exterior car parks should be lit in accordance with Section 4.13.

4.4.8.4   Airport-terminal buildings

Departure areas may well be lit to 500 lux, and it is normal practice to provide a lower level of lighting between there and the aircraft (on 'jetways'), to allow for visual adaptation along the route, and lighting designs should take account of this.

Diversity figures should be based on client specification but it is good practice not to exceed 0.1.

Care must be applied when designing the lighting outside terminal buildings, as these will be lit to significantly higher levels than the adjacent roadways. The illuminance values in Table B.5 of BS 5489-1 may not be high enough to ensure proper adaptation in these situations. Lighting designers should assess the requirements fully and, if necessary, increase the road lighting to Class CE0 as specified in EN 13201-2.

4.4.8.5   Passenger docks and wharfs

For safe movement, lighting of open quaysides should be to 50 lux average, 30 lux minimum. Covered areas should be treated as for bus stations (see above).

4.4.8.6   Bus stops

On-highway bus stops are mostly lit from the street lighting itself. Where the street lighting does not provide sufficient illuminance, the lamp rating of the nearest luminaire should be increased or a supplementary luminaire installed. Where bus-shelter manufacturers provide lighting within the shelter, this should be glare-free. If lighting is required at remote bus stops away from electricity supplies, it may be cost-effective to install solar or solar and wind-powered luminaires.

4.4.9 Recommendations

Table 4.5 Lighting recommendations for transport interchanges

Area to be illuminated	Environmental zone	Maintained average illuminance (in lux) not less than		Maintained minimum illuminance (in lux) not less than Horizontal value (E_h)	Uniformity ratio ($E_{h\,min}/E_{h\,ave}$) not less than	Colour rendering index (R_a) not less than
		Horizontal value (E_h)	Vertical value (E_v)			
Roads						
	E4	30		12	0.4	20
	E3	20		8	0.4	20
	E1 & E2	15		6	0.4	20
General pedestrian areas						
	E4	20		12	0.4	20
	E3	15		6	0.4	20
	E1 & E2	10		4	0.4	20
Information displays and advertisements			100[a]			
CCTV			10[b]			
Light rail systems						
Major boarding and alighting points and shelters				50		60
Intermediate boarding and alighting points and shelters	E4	25				60
Intermediate boarding and alighting points and shelters	E3	15				60

Table 4.5 (Continued)

Area to be illuminated	Environmental zone	Maintained average illuminance (in lux) not less than		Maintained minimum illuminance (in lux) not less than	Uniformity ratio ($E_{h,min}/E_{h,ave}$) not less than	Colour rendering index (R_a) not less than
		Horizontal value (E_h)	Vertical value (E_v)	Horizontal value (E_h)		
Bus and coach stations						
Bus stands		250			0.16	
Passenger-circulation areas		250			0.16	
Entrances, exits, amenity and waiting areas		250			0.16	
Ticket counters and information points		500^c				
Road/rail interchanges and railway stations						
Car park	E3 & E4	30			0.4	
	E1 & E2	15			0.4	
Covered car park	E3 & E4	75		50		
Drop-off points and frontages of stations	E3 & E4	40		20		
	E1 & E2	10		5		
Footbridges						
Open, level	E3 & E4	100		40		
	E1 & E2	50		20		
Covered, level	E3 & E4	150		60		
	E1 & E2	100		40		

Steps	E3 & E4	150	60	
	E1 & E2	100	40	
Subways, night				
Subways, day		150	60	
Open platforms	E3 & E4	50–80		0.4
	E1 & E2	30–50		0.4
Airport terminals				
Drop-off points and frontages of terminals		50	30	
Shipping				
Open passenger docks and wharfs		50	30	
Buses				
Bus stops	E4 & E3	20		0.4
	E2	10		0.4

Notes
[a] Working plane extends to area of display or advertisement.
[b] Up to 1.7 m above ground level.
[c] At desk top.

4.5   Effect lighting

4.5.1   Introduction

Effect lighting is a term used to describe the lighting of objects in the exterior landscape to introduce exciting or dramatic aspects or simply to draw attention to a particular feature.

Floodlighting is a technique that is widely used for effect lighting. Many floodlighting installations are purely functional, for instance security lighting for open areas where observation is needed at night, and as a deterrent to crime. Examples of functional floodlighting include sports floodlighting and safety lighting.

Floodlighting as effect lighting is different. It is carried out purely for aesthetics. The actual desired effect may vary from one project to another, but the objective is to make an impact or create a particular atmosphere. Primarily this is about lighting subjects, buildings, structures, sculpture, trees, tangible objects where light is used to enhance aspects of the subject.

Although the term 'floodlighting' is commonly used, it is probably incorrect to do so as it implies excess. The lighting should bring subjects to life after dark, rather than 'flooding them with light'. 'Effect lighting' is probably the nearest to an accurate expression, and the phrase is chosen both to declare the intention, and to use as a guide to what the object of the lighting is.

Effect lighting can equally mean designing shadow areas into the scene. The interplay of light and dark areas can play a major role in creating the finished view, and can add drama as well as revealing the nature of the structure. The generation of shadows can be introduced to accentuate structural details, making them a prominent part of the overall effect.

'Painting with light' is a phrase that aptly describes the way in which effect lighting can be used to the full. Painters think in terms of light and dark, form, colour, strength of detail and many other aspects of a subject in their presentation, and lighting designers should do no less.

Effect lighting can draw on psychological nuances which might be attributed to a particular scene, and be used to project the mood and nature of a building or structure.

All forms of lighting carry with them some engineering requirements in terms of design of lighting levels, positioning and mounting of equipment, electrical requirements and equipment design. Effect lighting, however, brings a strong component of aesthetic and artistic consideration; it is classically where the engineer meets the artist in creating a visual effect.

4.5.2   Scope

Probably the commonest form is the lighting of buildings. This is not surprising since there is a massive wealth of fine architecture, both new and old. The lighting is used to expose both the form of the building, and its finer details. By lighting carefully it is possible to give the building a 'new identity' after darkness falls. Colours can be enriched or changed.

Details not prominent under daylight can be emphasised. An important structure by day can become a major landmark by night. And the relationship of the building to its surroundings can be re-stated by the use of sympathetic treatment of the site.

The millennium celebrations in the United Kingdom included a great burst of activity in the use of effect lighting for Church buildings. Religious buildings are a large part of any nation's architectural heritage, and it is right that they should benefit from the attentions of good effect lighting.

Major structures such as bridges can also benefit enormously from carefully designed effect lighting. The complexity of the assembly or the elegance of the shape can be particularly emphasised when lit against a dark sky. A good example of how lighting can accentuate the beauty and shape of a bridge is shown in Figure 4.5.

Sculpture has not been so common in the public landscape, but this is changing with the introduction of new works being commissioned in many town and city centres. Colour can be vital, surface textures can be emphasised, but most importantly the dynamic movement captured in a

Figure 4.5 Millennium Bridge, Gateshead. (A stunning piece of architecture, achieved with innovative engineering, as it is almost changed into a sculpture by the use of lighting effects. The lighting under the bridge accentuates the curvature of the structure.) (*see Colour Plate 6*).

Figure 4.6 Ballerina, Covent Garden, London. (A single narrow beam floodlight mounted above the sculpture is aimed downwards to simulate stage lighting.) (see *Colour Plate 7*).

static object can be revealed to clearly show the sculptor's intention in the design. Some forms of sculpture are designed with lighting as an essential component of the piece. A typical illustration of a single narrow beam floodlight mounted above the sculpture is shown in Figure 4.6.

The combination of lighting effects and water features can also create a key feature, with the night-time appearance being transformed and a new dimension added to the scene. Colour can add even more to the effect. Specialised underwater lighting equipment is available to take account of the unique requirements of safety and maintenance. Fibre-optic systems offer interesting opportunities here because the light-generating equipment is separate from the delivery point and the normal concerns on safety and maintenance are eased. In particular, colour effects are easily achieved. Where fountain systems are included, it is often possible to synchronise colour change effects with water jet change programmes (Figure 4.7).

With a growing emphasis on the environment, it is not surprising that some attention has been paid to the lighting of natural subjects. Landscape

Figure 4.7 Fountains. (Fibre-optic lighting gives stunning effect to this mixture of water and sculpture.) (see *Colour Plate 8*).

architecture often pays great attention to trees and other green cover, but in all too many cases the view is lost after dark. Trees lit at night can be dramatic in their visual appeal, and careful effect lighting can soften the harshness of a winter nightscape. Disturbance of natural wildlife must be avoided, and consultation with local environmental authorities will help to define the needs here. Disruption of nesting sites for instance, could be avoided by seasonal adjustment to the operating times of the lighting arrangements.

4.5.3 Prime considerations

4.5.3.1 How to light?

There are examples around of buildings that have been 'flooded' with light, usually using projector floodlights operating from some distance. This has the effect of masking detail, and offering a bland appearance to the viewer.

Modern thinking is often to highlight only the details of a building, and to give a low-key light wash to the rest. This offers lower energy costs, and a much more effective revelation of the building as a piece of architecture. To really expose the character of a structure it is often good to design shadows into the view. The mix of lit and dark areas can be used to add mystery, and to strengthen the three-dimensional impact. You would therefore need a mixture of smaller units to light details (perhaps concealed in the structure), and others with a more general distribution to cover the broader areas. A good example of this would be in the lighting of church buildings. Heavy buttressing along a wall, or around the base of a tower can be emphasised to add a dimension of strength and solidity to the scene. The deep shadowing effect that can be produced will exaggerate the apparent size and depth of these surface features. Section 4.5.3.4 deals more fully with the use of directional lighting.

Where the effect required is to emphasise a particularly strong feature or line, then that item must be clearly identified in the design brief. In modern buildings there is often an exposure of frame detail, and some stunning effects can be achieved just by lighting that. For example, an exposed building frame can be shown in lit detail. The use of silhouette or shadow can add to the visual interest, and help to focus the eye on the architect's key design elements which can then become visual features in their own right after dark. Where detailed features are to be the key element in the night scene, then multiple small lighting fittings will need to be used to bring about the effect. Fixing fittings to exterior surfaces may be easy when the builder's scaffolding is in place, but a potentially difficult access at later times should be carefully considered during the design process. Equally restricting can be the requirements for listed buildings, where fixing to surfaces may not be permitted; again, this may limit the design options related to positioning of equipment.

4.5.3.2 How much light?

The most effective schemes often use comparatively little light. The important factor here is to balance the visual importance and appeal against adjacent buildings. Lighter surfaces will reflect more than darker types, and so, with the same illumination, will appear brighter to the viewer. Equally, heavily textured surfaces will absorb more light than smooth ones. So a dark, heavily textured surfaces will require substantially more light than a light smooth finish. It would also require more electrical energy.

Therefore it is recommended that the services of a lighting engineer are engaged to ensure the right amount of light is used.

4.5.3.3 Colour and colour effects

For any surface to retain its natural colour after dark, it must be lit with a light source which contains at least its own colour.

Colour filters are very useful. These can be a temporary add-on accessory for seasonal or festive events, or they can be used on a permanent basis where a sustained colour emphasis is required. A colour filter absorbs all colours except its own colour, e.g. a red filter, in the path of white light, transmits red light only, and blocks all the other colours. This must be taken into account in the planning since filtration means light loss, and perhaps a higher wattage lamp will be needed when colour filters are used. For example, a blue filter may transmit only 30 per cent of the light which is incident upon it, depending on the density of colour.

Colour can also be a key factor in giving a clear identity to a lit structure when other adjacent subjects are lit as well. The materials used in the structure may provide the inspiration for the colour, or the choice of lamp or the use of colour filters may add them in. The use of colour at night can dramatically alter the form, shape and character of an object as shown in Figures 4.8 and 4.9.

Light sources that are monochromatic, or strongly biased towards a small range of colours, will reveal only some of the colours resident in the structure. A 'full spectrum' lamp is preferable. Metal halide sources are excellent, having a white output, in revealing all the variations in the colours of brick or stone.

To some extent, the colour temperature of the light can be matched to the type of surface to be lit. For instance, yellow sandstone can benefit from the use of high pressure sodium lamps; these have a 'gold' appearance, and will emphasise the warm nature of the stone surface. Portland stone would best be served by metal halide lamps, which have a very white output. Fuller information on lamp types is found in Section 3.2.

4.5.3.4 Direction

Daylight has a generally downward bias. In direct sunlight this can be extreme, but even an overcast sky will form shadows below architectural details.

Floodlighting aimed upwards will reverse the shadows, and will give a different perspective to the structure. The closer the floodlight is to the surface, the stronger will be the shadows generated from surface detail.

Equally, it is often desirable to aim floodlights sideways, to direct light laterally across the surface to reveal the shadows from vertical features on the building, again enhancing the visual interest. Shallower aiming angles will again dramatise this effect more strongly. Care should be taken to avoid aiming angles in line with the main observer viewing positions in order that an unappealing 'flat' appearance is avoided.

Lighting from low level also makes the mechanics of installing equipment, and maintenance, that much simpler, but on the other hand may expose equipment to vandalism (refer to Section A.3.1). Where there is access for

Figure 4.8 'Slipstream' sculpture, South Lanarkshire.

Figure 4.9 'Slipstream' sculpture, South Lanarkshire. (Use of colour at night changes the character of the sculpture. As well as the illuminated and shadow areas.)

maintenance and suitable mounting locations, the opportunity should be taken to reduce light spill by illuminating from the top of the structure.

4.5.3.5 Close offset technique

Where sources are placed close to the surface to be illuminated, with the light directed across the surface, the technique is known as 'close offset' (see Figure 4.9), and is very effective in showing details of the building, not just the texture, but also accentuating the surface features.

4.5.3.6 Contrast

If a bland, uniform lit effect is to be avoided, then variations of levels of light on different parts of the subject will be needed.

Huge differences between the brightest and darkest areas will give a stark, heavily dramatised appearance. This will be particularly so if the contrasting areas are positioned close together. Extreme contrast will give an air of mystery, even 'fear', while moderated levels will soften the appearance. The human eye is remarkable in being able to deal with wide variations of lighting level, contrast and colour. It also measures distance, and adjusts for

Table 4.6 Effect of different contrasts

Contrast	Effect
1:1	Not noticeable
1:3	Just noticeable
1:5	Low drama
1:10	high drama

Note
These values are not suggested lighting levels, but
show the preferred ratios of low to high luminance
levels for typical projects. Experiments on site are
strongly advisable in determining contrast levels.

brightness levels. Even more remarkable is the ability to sense the 'mood' of lighting.

Too little contrast, on the other hand, will give a bland, over-soft appearance, and will probably not reveal the character of the subject sufficiently.

Of course there will be cases where a stark, angular, theatrical type effect is wanted.

The contrast levels that have been found to be good practice are shown in Table 4.6.

4.5.3.7 Beam control

Floodlights have a beam distribution that is largely dependent on the shape of the reflector and the positioning of the lamp within it.

A round reflector will produce a conical beam, symmetrical between vertical and horizontal axes (Figure 4.10). This 'symmetrical' type is useful for medium-distance requirements, typically to pick out important features (see Figure 3.20). Obviously a conical beam directed obliquely on to a surface will light an elliptical area. The centrally mounted lamp generates a conical beam.

For longer throws a 'narrow beam' symmetrical type is particularly useful (see Figure 3.19). The light output is packed into a narrow cone, with less spillage outside the target area, and the resulting higher intensity offers a long range unit to give sufficient illumination when it impacts on the surface. The shape of the reflector system concentrates the light into a very narrow cone.

A rectangular reflector, curved in one direction only (see Figure 3.21B) with the lamp placed centrally, will produce a beam which is asymmetrical between its two axes and useful for lighting areas rather than small points.

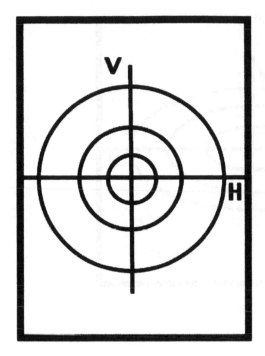

Figure 4.10 Symmetrical floodlight beam.

The single curvature reflector, and the centrally mounted lamp produce a beam roughly rectangular in section (Figure 4.11).

A rectangular reflector with different curvatures, and with the lamp offset (see Figure 3.22) will produce a distribution where the vertical output is offset, and also different from the horizontal axis output, this is termed as double asymmetric distribution.

The variation in the reflector profile and in the offset lamp position combine to give a strong accent on the upward part of the beam (Figure 4.12).

4.5.3.8 Accessories for beam control

It is likely that an 'off-the-peg' floodlight may not give the precision required. Constraints in mounting positions will often generate a need for a modified beam, and add-on accessories such as refractor glasses, which vary the beam shape, may be vital in getting the right result. Other accessories may be needed to give a cut-off at the beam edge to prevent light spill, and avoid the problems of light trespass and light pollution. It is important that the choice of equipment offers the facility of such additions and controls. See Section 2.3.7 for examples of accessories.

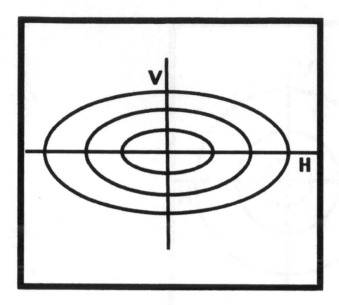

Figure 4.11 Asymmetrical floodlight beam.

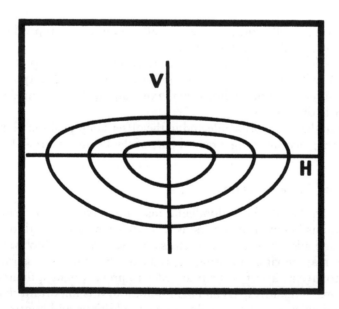

Figure 4.12 Double asymmetrical floodlight beam.

4.5.3.9 Daytime appearance

Equipment installed for night-time use will inevitably be visible by day. Where concealment is possible, for instance by using ground-recessed units, then the problem is overcome. There can, however, be problems when equipment is mounted on the surface of a structure, or other visible positions.

Obviously, using the smallest possible units can help here, particularly if partial concealment can be achieved behind some surface feature. An additional safeguard is to use equipment specially coloured to match the immediate surroundings, so that it does not intrude into the daytime view.

For localised lighting effects, fibre-optic systems can offer benefits. Fibre optics allow light to be generated from a remote position, and then delivered to one or many points of the subject. The emitting point is small, perhaps as small as 4.5 mm diameter, so the impact on daytime appearance is minimal. Fibre optics are also available as side emitting types, where the light is dispersed along the length of the fibre. These can be used to follow the lines of a feature to give emphasis to the overall form at night, whilst remaining unobtrusive by day. In such cases the fibres are typically 10–11 mm diameter.

Fibre optics systems generally use metal halide light sources, which give white light and excellent colour rendering. Additionally, fixed colour filters can be installed in the generator, and in some types a colour wheel can be included to give changing effects. Colour change effects can also be synchronised, using electronic controls to link generators. Fibre-optic systems are best utilised for localised lighting effects rather than the primary light source for an area, as it is difficult to deliver sufficient light for the latter.

Where concealment or camouflage is not possible, a clear, bold statement can be made, using the lighting fitting as a feature. On modern buildings for instance where steel frames or other structural details are exposed, there is no reason why lighting equipment cannot be added to emphasise the 'engineering' feel of the project.

The problem of a conflicting daytime appearance is particularly relevant to listed buildings, where any surface modification will be restricted by the appropriate authorities. In such cases discussions at an early stage are essential to ensure that the needs of all parties are met.

4.5.3.10 Anti-vandalism measures

Vandalism should be regarded as inevitable when equipment is exposed to public access, and measures to protect against it should be incorporated into the design. Ultimately there is no protection against the determined vandal, but some protection will normally discourage casual opportunist vandalism.

Where equipment is easily accessible, for instance surface mounted at low level, wire-mesh grills attached to the luminaires can offer some protection. Better protection can be provided by using totally enclosing wire-mesh cages with a minimum separation of 100 mm from the luminaires to be protected. Further guidance is given in Section A.3.1.

Concealing a luminaire in a secure enclosure with either a protective grille or toughened glass front will inevitably lead to some reduction in light output. Allowance must be made for this in the detailed design.

4.5.3.11 Maintenance

Maintenance and planning for maintenance is essential and reference should be made to Section 3.7.

4.5.3.12 Balancing the whole view

In the urban environment there is usually a substantial amount of exterior lighting in use. This will comprise road, security and amenity lighting, as well as effect lighting on other structures in the same scene. If effect lighting is being considered for a structure, then there is a need to balance the view presented to the observer. Without this balance the effect lighting can become lost in the total aspect, or be overstated to the point where any subtlety is lost.

The visual identity of a structure can be accommodated as follows:

- First, the structure may have such an outstanding and differentiated style that there will be no confusion with adjacent sites.
- Second, the natural colour of the subject may be such that it is clearly distinguished from others, or perhaps colour can be added by the choice of lamp type or colour filters.
- Third, there may be a particular feature that can be picked out to offer the viewer a focussed view.

All these options relate to the structure in comparison with other sites close by. Provided that there is sufficient differentiation, the matter of balance against ambient levels of lighting is the next hurdle. For a town-centre site, illuminance levels four times that required for a building in a relatively dark village may be required to achieve a similar visual impact.

Where effect lighting is being considered for a large town or city, there will ideally be a lighting strategy in place already (see Section 2.4). Such strategies are essential in ensuring that visual harmony is maintained, and particularly useful when a multiple-site project is planned over a longer period.

4.5.4 Techniques in design

4.5.4.1 Basic rules

When designing effect lighting the use of design software to predict the result is only part of the process. Such aids will give the calculated figures, but will not necessarily reveal what the illuminated object will actually look like. With effect lighting, the impact of what is being planned has to be checked, and the normal method is to use some sample equipment, with trailing cables, and to try out possible positions and equipment types. Only then is it possible to gauge the likely result, not only of lighting on the target areas, but also to assess the degree of spillage, and the extent to which additional beam controls will be needed. This process also allows some confirmation of the amount of light required to give a visual balance with surrounding features. At the same time it is possible to determine the optimum mounting positions for lighting equipment.

Close-offset technique has been mentioned already, but is now considered in more detail.

Figure 4.13 shows a vertical surface lit from ground level. The equipment position is shown, in close offset, that is positioned relatively close to the surface to be illuminated. The peak beam, that is the most intense part of the output, is aimed at a point high on the surface. The lighting produced will be a function of the inverse square law (Section A.1.1) and the cosine law (Section A.1.2).

Surface illuminance is reduced with increased aiming angle. The larger the aiming angle the longer the aiming distance, which will further reduce illuminance through the inverse square law. The light playing on the lower part of the vertical surface comes from a less intense part of the beam, but it travels less distance, and suffers less from the cosine effect.

With the right choice of beam characteristic it is possible to produce a reasonably uniform illuminance across the surface, if this is what is required. The main difference will be in the extent to which surface texture is revealed, as the light to the furthest point brushes closer over the surface than at lower levels. It may also be necessary to add supplementary floodlights to infill areas of shadow cast by features on the building such as recessed areas or balconies (Figure 4.14).

Moving the lighting equipment further away from the base of the structure (Figure 4.15) will reduce the cosine effect, but the aiming distance will increase. The result will be greater uniformity, but a loss of the texture

Figure 4.13 Close offset lighting.
Note: The cut-off of the lighting and shadows cast due to the balconies.

detail of the lit surface. There is also the risk of light spillage both beyond and into the building.

When lighting very tall structures it will generally be necessary to use multiple luminaires aimed at different parts of the structure to ensure an even and adequate level of illumination over the structure (Figure 4.16).

In lighting a complex form, such as a sculpture, the numeric analysis becomes even less useful. The objective here is to give a night-time expression of the sculptor's work. Tiny variations in the positioning of equipment can have an important effect on the final appearance. Colour may be even more significant, and the intensity of the projected light must be balanced against the possible variances in the nature of the lit surface. For instance, intense light on a bronze piece can result in ugly and distracting reflections from the surface.

Figure 4.14 Close offset lighting with additional luminaires added to infill shadowed areas.

4.5.4.2   Ground-recessed luminaires

4.5.4.2.1   GENERAL

Lighting from ground-recessed equipment is becoming very popular. Luminaires are now available to house different lamp types and wattages to suit many varied applications (see Section 3.4.2.2). The mounting position carries with it a number of attractive features and this type of luminaire is especially useful in close offset applications. There are, however, some important constraints, and these are outlined below.

Ground-recessed lighting can be used to light a wide variety of subjects. Apart from the lighting of objects such as sculpture and foliage, these units can also be used to identify pathways. This can give a reassuring feel to pedestrians, when their route options are clearly defined. LED clusters are particularly useful for route definition.

Figure 4.15 Lighting using luminaires situated further away from the structure.
Note: The reduction in the cut-off of the light due to the balconies.

However, ground-recessed luminaires do not give any horizontal illumination to the routes. The floodlighting of building façades is also a very real possibility with ground-recessed units. The higher powered versions are floodlights in the fullest sense, and can offer an effective night solution, plus an unobtrusive daytime appearance. Internal adjustments offer offset (angled) lighting direction, and external attachments can give a screened appearance to avoid any awareness of the lighting source.

Other prominent above-ground features can also benefit from this form of lighting. Trees, sculptures and other monuments can be given a dramatic night-time effect with the use of upward light. It introduces a dramatic contrast to daytime appearance, and can add a new dimension to the way in which the subject is seen. As with more conventional forms of floodlight, colour filters can be used to saturate the natural colours in the subject.

4.5.4.2.2 MECHANICAL ISSUES

Putting any electrical equipment into the ground potentially brings together the combination of electricity and moisture. It is therefore necessary to

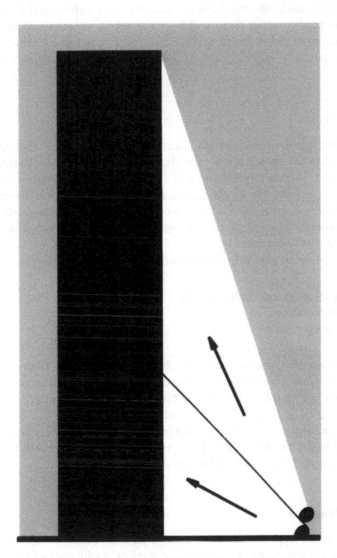

Figure 4.16 Lighting using multiple luminaires in a close offset configuration to illuminate a tall structure.

use equipment that has a minimum IP rating of IP67. Manufacturers of most ground-recessed units will specify drainage by use of a gravel layer to 300 mm under the unit, and also packed around its body, to minimise the proximity of water to the equipment. However, it is difficult to achieve this in a civil-engineering project.

Of equal importance is the need for the unit to carry substantial loads. Apart from pedestrian traffic, there will be areas where vehicles are used, and in any event most public areas will need access to emergency vehicles. A typical load capacity of 5000 kg or more will be required to accommodate this.

4.5.4.2.3 PUBLIC SAFETY

Luminaires generate heat as well as light, and this becomes an issue when the surface temperature of exposed areas exceeds levels of comfort and/or safety for human contact. Surface temperatures of 150 °C are not uncommon, but if the luminaire is beyond the direct access of the public then that should not present a problem. In many cases, however, luminaires are accessible, and it is then necessary to ensure that the surface temperature is within a safe range if injury is to be avoided.

In such circumstances, ground-recessed luminaires should have

- Low energy (fluorescent) light sources
- Built-in heat filters or
- A double-glazed visor.

'Dichroic' filters are the simplest way of achieving temperature limitation where high energy sources are used. These devices allow the transmission of light, but heat energy transmission is blocked, and is dispersed back through the fitting-body structure into the ground.

A surface temperature of up to 70 °C maximum is relatively safe, but care should be taken in the vicinity of water and where young children will have access (see Section A.3.1 for more information).

4.5.5 Water features

Water, like air, is transparent allowing light to pass through; in this way, we are able to see the bottom of a swimming pool or a still pool of water. Because light passes through water it is difficult to light it unless the water is moving or has an additive in to make it opaque. The movement of water causes ripples on the surface that stop the water acting as a mirror by breaking up any reflections. Also the movement causes the formation of tiny bubbles in the water which reflect light back allowing the water to be lit.

Care should be taken when trying to light water from above to ensure direct reflections of the light sources are not seen by observers. This can be difficult to achieve especially if the water can be observed from various viewing angles and it is for this reason that underwater luminaires are often used. These cast a light under the surface, lighting up objects without distracting surface reflections. Luminaires should be fully submersible and fitted with ample cable to allow them to be brought above the surface for maintenance. Alternatively, they may be installed in a purpose-built recess in the wall of the pool or river behind a transparent medium. This system allows easier maintenance but does restrict the locations of the luminaires to the sides of the water to be lit. All equipment and electrical wiring should be fully submersible and take account of the special requirements due to the potential corrosive nature of the surroundings.

The changing movement of water in a fountain or tumbling over a weir make an ideal feature to be illuminated by night. Water cascading over a weir is best lit from behind where the light will be seen as flickering and ever changing with the movement of the water. However, lighting up the water from below by aiming lights directly along the cascading water can also produce interesting movements in the water.

Fountains are best lit by directing narrow-beam submersible luminaires along the length of the water jet either from one end or from both, depending upon the length of the jet and its shape. The angle of the luminaires needs to be carefully established to ensure maximum utilisation of the light. Such luminaires whilst having to be submersible will need to be of substantial construction with adequate locking mechanisms to stop them being moved by the force of the water playing on them.

Further interest and drama can be added to water features by the use of colour changes. This can be achieved by the use of different light sources or luminaires fitted with colour filters being switched on and off or by the use of fibre optics fitted with colour-change wheels. The reduction in the light output from the luminaires should be taken into account when using colour filters and colour-change wheels.

Table 4.7 gives an indication of the luminous intensity required to illuminate water features of varying heights.

4.5.6 Assessment of completed projects

As in any project, when the installation is complete, there is an opportunity to review the finished result. In effect lighting, because the requirement is to produce a visual statement, this assessment becomes essential and may lead to some alterations after completion.

The checklist for assessments is as follows:

- Does the lighting achieve the required effect?
- Are the important features properly illuminated?

Table 4.7 Luminous intensity required to illuminate
water features of varying heights

Height of water feature (m)	Total luminous intensity of source at base of fountain (cd)
1.5	4 000
3.0	11 000
6.0	34 000
9.0	69 000
12.0	115 000
15.0	170 000

- Is the character of the structure properly emphasised? (i.e. are any changes needed to the aiming?)
- Is the brightness level correct?
- How does it appear against its immediate environment (background, other features nearby, etc.)? (Particularly important if the project is to fit in with a longer-term lighting strategy.)
- Is the colour content correct?
- Is the scheme producing excessive levels of light pollution and, if so, are additional attachments required to correct this? (see Section 2.3.6).

4.5.7 Summary

This chapter started out by stating that effect lighting is different from other forms of exterior lighting. It brings together all the techniques of lighting, focussed on producing a lit effect. It is where the artist joins hands with the lighting engineer, in pursuit of a visual statement. Huge opportunities are presented for the expression of form, colour, contrast, and design – all the things that add an aesthetic dimension to the environment.

Effect lighting can have an enormous effect in adding both visual interest, and in expressing the character of structures, towns, and cities. Such lighting should be seen as a beneficial investment in society today.

4.5.8 Recommendations

Table 4.8 illustrates that a subject in a major city will need about three times as much light as the same subject in a country village, to be effective against typical ambient levels. In a project with mixed surfaces, a section in granite will need almost four times as much light as an adjacent area in white brick if the same apparent brightness is required.

Table 4.8 Suggested trial illuminance for effect lighting schemes

Material	Approx. reflectance	Surface condition	Suggested trial average illuminance (in lux) Environmental zone			
			E1	E2	E3	E4
White brick	0.8	Clean/new	5	15	25	40
		Fairly clean	10	60	90	180
Portland stone	0.6	Clean/new	10	20	35	60
		Fairly clean	15	60	90	180
Middle stone	0.4	Clean/new	15	30	50	80
Medium concrete		Fairly clean	25	100	150	300
Dark stone	0.3	Clean/new	25	40	60	100
		Fairly clean	30	200	300	600
Granite	0.2	Clean/new	25	55	90	150
Red brick		Fairly clean	40	180	270	540

Notes
1 The illumination levels given above are a guide for the initial design work. These should be verified by site visit and trial.
2 The reflectance factors given are for white light. Light content with a strong colour bias may need variation in intensity to compensate. This should be established by site trials.
3 No values are given for dirty buildings as it is considered that a cost–benefit analysis should be carried out to assess the benefit of cleaning the building as against the additional cost of providing and operating substantially increased levels of effect lighting.

4.6 Exterior work areas

4.6.1 Introduction

The purpose of exterior work area lighting is to produce a safe working environment and to provide adequate lighting levels for users' differing activities.

4.6.2 General considerations

Factors that will influence choice when designing a good lighting solution are as follows.

4.6.2.1 Selecting appropriate light source and wattage

When selecting a suitable light source for a particular application, the main determining factors will be the luminous efficacy of the lamp and its colour rendering abilities. High pressure sodium sources are used widely as they offer a good compromise between these factors. In certain applications a higher degree of colour rendering may be necessary – use of white light will dramatically increase the impact of the lighting, but, conversely, a lower lumen output and lumen depreciation factor will be expected which will increase the quantity of luminaires.

In order to minimise the impact of glare it is also necessary to use a lamp wattage corresponding to the luminaire mounting height. Table 4.9 shows maximum recommended wattage values for typical area lighting mounting heights. The quality of the lighting design can be further improved by achieving the overall uniformity shown in the recommendations (Section 4.6.11).

4.6.2.2 Selecting suitable luminaires

The performance of luminaires will be affected by the relevant pollution category; therefore a luminaire with an optic with a high degree of IP will operate more effectively. It will allow the use of higher maintenance factors thus achieving the same lighting solution with fewer luminaires. This has the double benefit of reduced installation and running costs.

Other issues like corrosion resistance may also need to be considered, particularly in areas that are near to the coast.

4.6.2.3 Maintenance and re-lamping

Maintenance is another important feature when looking at the overall design. The luminaires will need to be regularly accessed in order to facilitate cleaning or re-lamping. This is usually achieved with raise-and-lower platforms, which need to be able to reach the luminaires themselves. In areas where vehicle access to columns is restricted, it may be feasible to install raise-and-lower columns in order to access the luminaires at ground level. High-mast lighting units will usually have a cradle that can be lowered to ground level for maintenance, and a 4 m-by-4 m working area should be provided.

Table 4.9 Maximum recommended wattage

Typical mounting height (m)	Maximum recommended lamp wattages (W)
8	150
10	250
12	400
15	600
>15	1000

Note
Value for mounting height for exterior work areas.

4.6.2.4 Awareness of light spill

Light spill should be considered as an integral part of the lighting design process. Better control of lighting is required in order to reduce the amount of spill light around the perimeters of illuminated exterior work areas. Careful aiming of the luminaires can dramatically reduce direct upward light. Better understanding of luminaire photometry will increase the levels of useful light and decrease the light spill. Where control of light is necessary to ensure light does not extend beyond the area required, the use of louvres is recommended. Further information is given in Section 2.3.

4.6.3 Building sites

The main requirements of building-site lighting are as follows:

- Good vertical illuminance, as well as horizontal illuminance, may be required to enable workers to perform certain activities more effectively.
- Glare must be avoided as it could increase the potential of an accident occurring on a building site.

A particular feature that makes a building site environment hazardous is the continual changing work landscape. Combined with the plant, equipment, and building supplies, good lighting solutions are necessary to ensure that the working environment is safe – this is especially important during the winter months.

Lighting needs to be installed in areas where vehicles and luminaires will not be in conflict, and ideally not in areas for future construction. Additional lighting may also be required to aid workers performing certain tasks. A benefit of using portable supplementary lighting equipment would be that it can be re-positioned as construction progresses.

Another aspect of building-site lighting is the requirement for site security. Building sites can be prone to vandalism and theft – the addition of perimeter lighting can reduce the possibility of such an occurrence. This lighting may only be required outside of working hours during the night to reduce the overall running costs.

4.6.4 Rail – sidings/marshalling yards/goods depots

The following points should be considered when lighting these areas:

- Confusion of lighting with signalling – a higher mounting height and regular luminaire arrangement will reduce the chance of confusing signalling and lighting.

- Reflections from signals – ensuring luminaires are carefully positioned will help minimise these reflections.
- Glare can be minimised by using an appropriate lamp wattage for its corresponding mounting height, and by careful placement and aiming of the luminaires.
- Electrical safety – care must be taken to ensure columns and luminaires are not in close proximity to electrified lines.
- Vandalism – given the nature of railway environments, luminaires, particularly when easily accessible, should be of robust construction with tamper proof fixings.

Visibility can be improved by trying to ensure that the luminaires, railway track, and observer are co-linear (Figure 4.17). If this arrangement is not suitable it is possible to aim the luminaires perpendicular to the direction of the track. However, the effectiveness of the lighting will be reduced because of shadowing caused by the wagons. Design must account for these shadow effects. Vehicles must be lit below solebar level for maintenance examination/train preparation purposes.

Careful selection of the luminaire or lamp position is also required to ensure 'glare' will not cause a problem to drivers or nearby workers. This can be achieved by ensuring the peak beam is located below an angle of 70° from the vertical.

As a rule of thumb, the maximum column spacing will be approximately equal to twice the luminaire mounting height – e.g. a typical spacing for a 12-m mounting height would be approximately 24 m. This will obviously vary from scheme to scheme depending on the local conditions.

A siding is used to temporarily hold wagons that are waiting to enter a marshalling yard for sorting. In the marshalling yard, itself, the wagons pass through the switching area or 'neck' for sorting into its correct wagon destination. The lighting of the 'neck' is higher than for sidings in order to improve the visibility of moving wagons and to detect possible track obstacles. A goods depot primarily consists of sidings to transfer containers from road to rail, and sidings for longer-term storage of wagons. Similar to the 'neck' in a marshalling yard, the greatest level of illuminance is required in the loading/unloading area where containers will need to be moved.

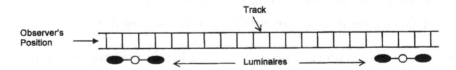

Figure 4.17 Luminaire orientations with regard to rail track.

4.6.5   Sea – dockyards/container terminals/jetties

The main requirements of lighting of dockyards, and related facilities, is similar to that of marshalling yards. The main areas of concern are as follows:

- Glare must be avoided to passing shipping and ships docking in adjacent bays. Careful aiming and location of luminaires can achieve this.
- Confusion of lighting with navigational equipment – a higher mounting height and careful luminaire location will avoid this problem.

Container terminals involve the transfer of containers from rail or road onto large container ships. The areas that require a higher level of illumination are those where containers are being moved – these include the jetties where containers are loaded onto ships, and areas where containers are loaded onto lorries or trains (Figure 4.18).

A high-mast lighting system may be more suited to this particular application. The advantage of using a high-mast system is the fact that the number of lighting masts will be reduced, which will help minimise the physical impact of the lighting system. Another important factor is that in areas

Figure 4.18 Lighting of container terminal (see Colour Plate 9).

where containers are stacked on top of each other, shadowing will cause problems that can be greatly reduced by increasing the luminaire mounting height.

Care should be taken to ensure there is a transition zone between the higher illuminance and lower illuminance areas to help workers adapt more effectively to the changing lighting levels.

4.6.6   Sales areas – car forecourts

Lighting used in sales areas such as car forecourts needs careful consideration, due mainly to the product being sold. The main areas of concern are as follows:

- Choice of light source – the colour a car appears depends on the spectral composition of the light and the way the colours within the light are reflected, therefore the light source used can change/influence the customer's perception and affect their choice of car.
- Security – cars to a potential criminal are very inviting due to their value; good lighting can deter and prevent the criminal from feeling safe.
- Light pollution – due to the increasing number of business parks appearing on rural sites, the control of obtrusive light is becoming more important. Obtrusive light impedes our view of the sky and causes what is commonly known as 'sky glow'. Choosing luminaires that prevent light above the horizontal can reduce this.

A natural/white light source should be used to allow the cars to appear as they would during daylight. For example, other light sources such as low pressure sodium produce virtually pure yellow light, which would turn a red car in daylight into a brown car at night.

Security is a major factor especially as most forecourts are unoccupied at night. The main areas to be lit are the perimeter of the site and the areas around barriers and entrances, ensuring that there are no dark spots in which a criminal could hide. The main showroom/building should be lit so that any intruder is seen in silhouette or their shadow is projected up onto the building.

Car forecourts should also be lit to standard car park lighting levels (see Section 4.13) to aid in the movement and parking of vehicles.

4.6.7   Coal mining – open cast

Lighting used in open cast coal mining is very different when compared to other areas due to the fact that coal has very little reflectance and normal lighting methods are not practical.

The main areas of concern when lighting open cast coal mines are as follows:

- Cost – due to the coal having very low reflectance it is uneconomical to light the area by conventional means.
- Limited illumination from vehicles – mounds of coal are levelled using trucks which use headlights to illuminate the surrounding area. This provides small pockets of inadequate, localised illumination.

Open-cast coal mining involves retrieving the coal from the ground ready to use as fuel for homes and businesses. Since more coal is used during winter, stocks of coal are built up during the summer, creating vast deposits of coal. Mounds of coal are created by heavy machinery such as bulldozers, which move the coal from the tippers onto the mound and also level off the coal.

Conventional lighting methods are unpractical due to the coal having very little reflectance, so a different method should be utilised. This involves creating shadows and silhouettes to reveal objects, i.e. hills and troughs of the coal mounds. Typically, to create these, floodlights are placed around the bottom and at the peak of the mound. An average illuminance of >5 lux should be achieved.

Areas where tasks such as loading, unloading, moving, and processing of the coal are carried out will require additional luminaires. Factors for consideration are similar to those in Section 4.6.2.

4.6.8 Loading bays

Lighting used in loading bays needs careful consideration to ensure the light reaches the areas of importance. The main areas of concern are as follows:

- Wasted illumination – luminaires positioned in the wrong position, such as in the middle of a loading bay, will provide little illumination due to the fact that the emitted light will be blocked by the vehicle loading.
- Height of loading bay canopy – the height of the loading bay canopy will affect the choice of luminaires used. Canopies at a height above 6 m will allow the use of HID light sources as well as fluorescent, which should be used for canopies below 6 m.
- Interior lighting of loading vehicle – luminaires positioned towards the inside of the vehicle allow loading to be carried out with a greater ease, and with higher safety.
- Uniformity – adjacent areas such as the inside of stores should be lit to no more than 10 times the loading bay illuminance.

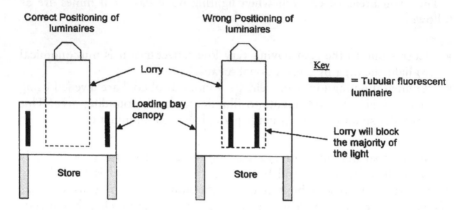

Figure 4.19 Position of luminaires on loading bays.

The positions of the luminaires within the loading bay are very important. One of their main purposes is to aid the driver in positioning the lorry to enable efficient loading of the vehicle. This is achieved by ensuring both sides of the vehicle are well lit and that the parked up lorry will not block any of the light emitted from the luminaires, which will cause shadows (Figure 4.19). Lighting the interior of the lorry being loaded is also important and can be achieved by attaching luminaires to the bay doors; these should automatically switch on when the bay doors open.

When high pressure sodium light sources are used glare can be a problem. To help eliminate glare, luminaires should be inclined/positioned so that the peak beam is outside the driver's line of sight. One way of achieving this is to position the luminaires to face backwards away from the driver; this though restricts the amount of light beyond the canopy edge.

4.6.9 Sewage/water treatment works

Lighting used in sewage/water treatment works needs to be carefully thought out to enable work to be carried out safely. The main areas of concern are as follows:

- Recognition of tanks – due to the substances being used, some of which are dangerous, identification of tanks is highly important.
- Shadows – due to tanks being in close proximity, and the height of the tanks, shadows can easily be formed. Dark areas need to be eliminated especially on walkways and where work is carried out.

Sewage/water treatment works are in large open spaces and the main work carried out is maintenance orientated. This work is carried out at the base of tanks, and requires adequate lighting to allow these operations to be carried out safely.

Columns need to be positioned between tanks ensuring the maximum utilisation of the light available. In situations where there are a number of tanks, mounting off the side of the tank should be incorporated to help reduce dark areas/shadows.

4.6.10   Petrol stations

Lighting used in petrol stations needs to attract customers while providing a safe area to fill up their vehicle. The main areas of concern are as follows:

- Identification – the products and services available need to be easily identifiable, so that the consumer can enter and position their car in the required position.
- Drawing the customer in – the appearance of the forecourt is a major factor in attracting customers. Attractive lighting of signs, advertisements, pumps, etc. create an impression of a quality filling station.
- Safety – the lighting should show a safe route into and out off the petrol station, with clear, easily identifiable signs.
- Method of lighting – recessed lighting within the canopy or up lighting using floodlights, where light is reflected of the underside of the canopy, will affect the way the petrol station is designed.
- Environment – obtrusive light should be kept to a minimum and all lighting should be in harmony with its environment.

The need to attract custom is vital to stay in business. Lighting is a large factor in the attractiveness and can invite customers to use the petrol station. Lighting should also increase the security, illuminate the pump and allow under-the-bonnet tasks to be carried out efficiently and safely.

The forecourt is lit usually from a canopy over the pumps. The lighting needs to make the user feel safe, identify the products available and make pay areas etc. recognisable. The two most common ways are to attach down lights to the underside of the canopy itself or fix up lights to the canopies supporting structure. When using the latter method, the underside of the canopy should be matt white and cleaned regularly to maintain maximum reflectance.

The advantage though is easy maintenance and installation, plus the canopy only needs to support itself and not the weight of the luminaires.

4.6.11 Recommendations

Table 4.10 Recommended lighting levels for exterior work areas

Areas to be lit. operations performed	Maintained average illuminance (in lux) not less than ($E_{h\ ave}$)	Maintained minimum illuminance (in lux) not less than ($E_{h\ min}$)	Overall uniformity not less than ($E_{h\ min}/E_{h\ ave}$)	Minimum colour rendering indices (R_a)
Building sites				
Very rough work, e.g. clearance, excavation and loading ground	20		0.25	20
Rough work, e.g. drain pipes mounting, transport. auxiliary and storage tasks	50		0.40	20
Accurate work, e.g. framework element mounting, light reinforcement work. wooden mould and framework mounting, electric. piping and cabling	100		0.40	40
Fine work, e.g. element jointing, demanding electricity, machine and pipe mountings	200		0.50	40
Traffic area				
Pedestrian passages. vehicle turning. loading and unloading points	50		0.40	20
Safety and security				
General lighting on building site, element mould, timber and steel storage. building foundation hole and working areas on sides of the hole	50		0.40	20
Pedestrian routes	50		0.40	20
General working areas	10		0.40	20
General storage areas	50		0.40	20
Rail – sidings/marshalling yards/goods depots				
Storage areas	10		0.40	20
Cleaning and service areas	50		0.40	20

Fuelling/effluent/washing points	100		0.40	20
Shunting neck	20		0.40	20
Turntables	100		0.40	20
Switching areas	20		0.40	20
Sorting sidings	10		0.40	20
General yard areas	10		0.40	20
Authorised walking way	10	5		20
Note: All values with rolling stock in position.				
Sea – dockyards/container terminals/jetties				
Container and bulk cargo terminals	50		0.40	20
Loading/unloading areas	50		0.40	20
Stacking areas	10		0.40	20
Jetties	50		0.40	20
Quayside	50		0.40	20
Sales areas – car forecourts				
Rural E1 and E2 (low district brightness)	15	5	0.20	≥65
Urban E3 and E4 (medium/high district brightness)	30	10	0.20	≥65
Coal mining – open cast				
Coal mounds	5			20
Loading bays				
Loading bays	150			20
Petrol station				
Access/approach area leading to the forecourt	50			20
Forecourt and pump areas	300			≥65

4.7　Hazardous areas

4.7.1　Introduction

Legislation in UK and EC requires owners to ensure that the operational procedures of their plant, whether these be large offshore oil installations, petrochemical complexes, small garage forecourts, distilleries, paint shops, inspection pits or sewage works, guarantees the safety of the workforce and the public. The prime requirement therefore is to establish local and operational conditions that obviate the creation of a potentially explosive atmosphere.

This section deals with technical aspects of outdoor lighting design and installation where the area being illuminated is designated as a 'hazardous area'. Safety protection specified for electrical lighting equipment in a designated hazardous area must be appropriate to the environment in which it will operate. Fitness for function together with an appropriate level of safety protection are the primary priorities. The integrity of all electrical lighting systems designed for use in hazardous areas is dependent upon the system being correctly installed inspected, tested, operated and maintained in full compliance with engineering regulations and standards.

The following information should not be construed as a replacement for operational procedures given in regulatory documents. The section is broad based so as to present an overview of the standards and principles underpinning the lighting of outdoor hazardous areas; it is not designed to be comprehensive; and its intention is to heighten awareness, and reinforce and complement information from sources such as the Department of Trade and Industry (DTI) and the Health and Safety Executive (HSE). When specific details and operational requirements are sought, engineers should refer to source documentation.

4.7.2　Hazardous area classifications

4.7.2.1　Explosions

For an explosion to occur, three components must feature. First, there has to be a source of chemical energy. The chemical energy is provided by the presence of fuel that can exist in the form of a gas or dust. Second, the presence of oxygen is needed to enable the fuel to burn. Third, there has to be a means of triggering the combustion. This can be achieved by the presence of an ignition source such as a hot surface or a spark.

For a mixture of a combustible gas/dust and oxygen there is a critical concentration, called the most easily ignited concentration, where a minimum critical quantity of energy will cause the mixture to ignite. If less than that critical amount of energy is released into the mixture, a self-propagating explosion will not occur. Some combustion may occur transiently, but a combustion wave does not grow and become self-propagating. However, if the critical amount of energy or more is delivered to the mixture of fuel and oxygen, the combustion wave will pass through the incipient stages of growth and become self-propagating, resulting in an explosion.

4.7.2.2 Explosive and potentially explosive atmospheres

An explosive atmosphere is one where flammable materials in the form of gases, vapours, mists or dusts are mixed with air. If an explosive atmosphere is ignited, combustion will take place and the combustion will affect the entire mixture. A potentially explosive atmosphere is one which would explode if certain local and operational conditions prevail.

4.7.2.3 Zonal classification

The degree of the risk presented by a hazardous area is dependent on the probability of finding dangerous quantities of flammable substances in the atmosphere of the area. Hazardous areas are classified in zones in accordance with the nature of flammable material that may exist within the area and the risk the material presents by way of causing an explosion.

Table 4.11 outlines the zonal classification for hazardous areas. The criteria for the zones are only a guide. The classification for an area can be affected by other factors such as, for example, the degree of ventilation.

4.7.2.4 Gas groups

Gases that could create a potentially explosive environment are classified under two groups, Group I and Group II.

Group I is concerned with gases that relate to underground applications associated with the mining industry. Group I comprises methane (firedamp) gas.

Table 4.11 Criteria for zone classification of hazardous area

Classification of hazardous area	Criteria for the zone
Zone 0 (gases) Zone 20 (dusts)	Flammable material present continuously for long periods
Zone 1 (gases) Zone 21 (dusts)	Flammable material present in normal operation
Zone 2 (gases) Zone 22 (dusts)	Flammable material present in abnormal conditions only for short periods

Gas Group II comprises those gases relating to surface industries. Gas Group II is divided into subgroups IIA, IIB and IIC according to their ability to be ignited by the spark produced by a discharging capacitor. In Group II, the gases within IIC gases are the easiest to ignite. Hydrogen gas and other gases requiring similar ignition energy (around 40 mJ) comprise subgroup IIC. Subgroup IIB comprises ethylene and other gases requiring ignition energy of around 160 mJ. Subgroup IIA comprises propane and other gases requiring ignition energy of around 300 mJ. Some common flammable materials relating to the subgroups of Group II are shown in the Table 4.12.

Equipment designed for use in an atmosphere that is likely to involve gases from subgroup IIC would meet requirements for IIB and IIA. Equipment designed for use with IIB meets the requirements for Group IIA. Equipment that is certified for use as Gas Group II is suitable for use with subgroups IIA or IIB or IIC. The use of equipment is confined to specific environments where the degree of risk presented would not warrant the use of a higher specification and as a consequence over specified equipment, which would increase costs.

4.7.2.5   Ignition temperature classification – gases and dusts

In addition to the energy transferred by a spark, gases may also be ignited through contact with a hot surface. The ignition temperature of a gas is defined as the temperature of a hot surface in contact with the gas that will

Table 4.12 Subgroup classification of common flammable Group II gases

Common flammable materials in Group II		
IIA	IIB	IIC
Acetic acid	Ethylene	Acetylene
Acetone	Propan-1-ol (n-propyl alcohol)	Hydrogen
Ammonia	Tetrahydrofuran (THF)	Carbon disulphide
Butane	Cyclopropane	
Cyclohexane	Butadine	
Ethanol (ethyl alcohol)	Dioxin	
Kerosene		
Methane (non mining)		
Methanol (methyl alcohol)		
Methyl ethyl ketone (MEK)		
Propane		
Propan-2-ol (iso-propyl alcohol)		
Toluene		
Xylene		

cause the gas to ignite and sustain combustion. The gases within Group I and subgroups IIA, IIB and IIC are further classified in terms of their ignition temperatures. The gases are assigned what is known as a 'Temperature Rating' or 'T rating'. The relationship between the T rating, the ignition temperature range for the gas and the corresponding maximum surface temperature for safe use is shown in Table 4.13.

Town gas, for example, will ignite on contact with a surface at a temperature of 600 °C. This gives it a temperature rating of T1. Carbon disulphide vapour has an ignition temperature of 100 °C, so it has a temperature rating of T6. Obviously, it is important that the surface temperature of equipment that is to be used in a hazardous area will not exceed the ignition temperature of the gas that may be prevalent in the area. Therefore equipment approved for use in an area where the atmosphere had a T6 rating would assure that the maximum surface temperature of the equipment did not exceed 85 °C.

The T ratings for some of the common materials in subgroups IIA, IIB and IIC are given in Table 4.14. As well as explosive gases, it may be that equipment will be operated in environments where explosive dusts are present. Where this is the case it will be important to know the ignition temperature of the dust concerned, both when it is in the form of a cloud and when it forms a layer. Layers of dust collecting on the surface of an appliance can produce an insulating effect, thereby reducing the rate at which heat is dissipated and causing the surface temperature of the appliance to increase above the value quoted for gaseous environments. Manufactures are required to account for this phenomenon in specifying their equipment. In addition to quoting the T rating for the appliance, the manufacturer will also quote a value for the maximum surface temperature of the appliance operating under dust-layer conditions. Ignition temperatures for some dusts (metallic and non-metallic materials) in both cloud and layer form are shown in Table 4.15.

Table 4.13 Relationship between the T rating, the ignition temperature range for a gas and the corresponding maximum surface temperature for safe use

Temperature rating (T rating)	Ignition temperature range of gas (°C)	Maximum surface temperature for safe use (°C)
T1	greater than 450	450
T2	between 300 and 450	300
T3	between 200 and 300	200
T4	between 135 and 200	135
T5	between 100 and 135	100
T6	between 85 and 100	85

Table 4.14 'T' rating for common materials

Subgroup	Material	T rating
IIA	Acetic acid	TI
	Acetone	TI
	Ammonia	TI
	Methane (non-mining)	TI
	Methanol (methyl alcohol)	TI
	Methyl ethyl ketone (MEK)	TI
	Propane	TI
	Toluene	TI
	Xylene	TI
	Butane	T2
	Ethanol (ethyl alcohol)	T2
	Propan-2-ol (iso-propyl alcohol)	T2
	Cyclohexane	T3
	Kerosene	T3
IIB	Ethylene	T2
	Propan-1-ol (n-propyl alcohol)	T2
	Tetrahydrofuran (THF)	T3
IIC	Hydrogen	TI
	Carbon disulphide	T6

Table 4.15 Ignition temperature for dust material

Dust material	Ignition temperature (in°C)	
	Cloud	Layer
Aluminium	590	>450
Coal dust (lignite)	380	225
Flour	490	340
Grain dust	510	300
Phenolic resin	530	>450
PVC	700	>450
Soot	810	570
Starch	460	435
Sugar	490	460

4.7.3 Standards

4.7.3.1 General

The need for transparent, harmonised standards is obvious. If free movement of electrical equipment across Europe is to be achieved then consumers need to be confident that the integrity of their electrical equipment is assured and has been tested and certified uniformly against common standards, irrespective of the place of manufacture.

4.7.3.2   The ATEX Directive and essential Health and Safety requirements

Essential Health and Safety requirements (EHSR) are embedded within a European Community directive that came into force in March 1996. This directive is commonly referred to as the ATEX directive. ATEX is an acronym derived from the French term 'atmosphère explosible'. The EHSR within the ATEX directive describe the criteria that must be met by equipment and protective systems that are intended for use in potentially explosive atmospheres.

The ATEX directive ran in parallel with other health and safety provisions that relate to the standards that lighting equipment must conform to in order to assure safe operation in potentially explosive atmospheres. The transition period during which these parallel arrangements applied ended on 1 July 2003. From that date all lighting systems intended for use in potentially explosive environments must comply with the EHSR as defined by the ATEX directive. The ATEX directive will facilitate the free movement of lighting goods within the European Community and ensure that the free movement involves goods that conform to a harmonised system of European standards.

The standards of the ATEX directive subsume the current European harmonised standards for electrical equipment. These standards, the so-called CENELEC standards for electrical equipment, which have been developed and refined over the years in line with technological advances in lighting system design, will largely cover the detail of the EHSR of the ATEX directive (Figure 4.20).

The main additional requirement to be met under the ATEX directive is the way in which the goods are certificated, marked and monitored. The markings on electrical equipment must include legible and indelible notification of the name and address of the manufacturer along with the specific marking of explosion protection and the Group and Category (see Section 4.7.3.4) to which the equipment belongs. Adherence to an ATEX specification for the documentation to be supplied by the manufacturer covering installation and operation of the equipment is also necessary. The requirements for the continuous monitoring of products are addressed within the ATEX directive and thereby ensuring surveillance of equipment will be undertaken in a uniform fashion by the appropriate authorities of the member states.

4.7.3.3   Conforming with standards

In order to demonstrate that a product meets the EHSR of the ATEX directive, it must undergo testing and receive a certificate of conformity from a Notified Body. A Notified Body is a government-approved independent

ATEX rating plate example

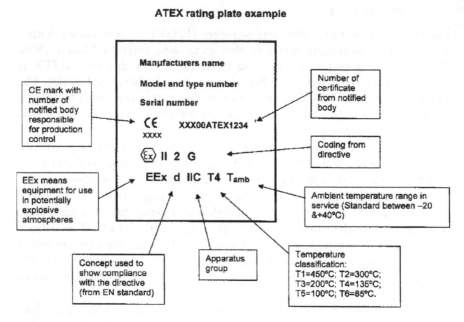

Figure 4.20 Typical equipment marking (ATEX). (Reproduced by permission of TRL.)

testing authority. In the UK, Notified Bodies are Electrical Equipment Certification Service (EECS), British Standards Institution (BSI), Transport Research Laboratory (TRL) and Sira Certification Service (SCS). Other members of the Community have their own Notified Bodies. The Notified Body scrutinises the documentation of a manufacturer's product and tests and evaluates the explosion protection properties of the product. Testing which shows that the product complies with the standards subsumed by the EHSR leads to the issue of a certificate of conformity by the Notified Body. The manufacturer can then mark the product with the CE label and, where certified to do so, also label it with the specific marking of explosion protection. The ATEX labelling denotes that the product can be sold anywhere within Europe without the need for further controls.

4.7.3.4 Equipment groupings and categories

Under the ATEX directive, equipment for use in potentially explosive atmospheres is divided into two groups, Group I equipment and Group II equipment.

Group I equipment is highly specialised, designed and constructed for use in the mining environment and therefore must safeguard against the presence of methane gas, firedamp. Within Group I there are two categories of equipment, Category M1 and Category M2. The specification for Category M1 includes the requirement that the safety provided by the equipment is such that in the event of failure of one means of protection a second independent means will provide the requisite overall level of protection. Category M1 equipment must have the facility to remain functional when an explosive atmosphere is present. Category M2 equipment must ensure safe operation even under rough handling and changing environmental conditions. Category M2 equipment is intended to be de-energised in the event of the occurrence of an explosive atmosphere. Full details of the requirements applicable to Group I equipment are included in Annex C of the ATEX directives.

Group II equipment relates to non-mining applications. Group II equipment is designed for operation in explosive atmospheres that do not involve methane gas. There are three categories of Group II equipment. The categories are labelled Category 1, Category 2 and Category 3. Full details of the EHSR for Group II equipment are provided in Annex C of the ATEX directive. Table 4.16 gives a general indication of the level of protection provided by the Group II categories and the zones (relating to gas and dust) in which the equipment can be operated safely.

4.7.4 Types of protection for lighting systems

It is important that the design of the system, the material from which it is constructed and its mode of operation preclude any possibility of the system acting as an ignition agent for the explosive atmosphere. In other words it is vital that the equipment is fully protected against itself.

A range of approved types of protection are used in the design and construction of lighting systems in order to ensure that the EHSR are met. Hot surfaces, sparks, arcs and electrostatic discharges are all potentially ignition sources. Electrical equipment for use in hazardous areas is designed to exclude, suppress or contain the effect of these sources. Types of protection include flameproofing the enclosure whereby, although the flammable atmosphere can penetrate the enclosure, the design contains the explosion, and its transmission outside the enclosure is prevented. This type of explosion protection is demoted by the code symbol 'd'. Another method encapsulating the fitting ensures that the explosive atmosphere is excluded and cannot interface with ignition components. This method is denoted by the symbol 'm'.

Different types of protection are appropriate to different zones. Types of protection for lighting systems for use in hazardous areas, their symbols, the

Table 4.16 General indication of the level of protection provided by the Group II categories and zones

ATEX Group II category	Level of protection provided	Safety requirements met (for full details see Annex C of the ATEX directive)	Intended operational area of the equipment	Corresponding zonal classification for gases and dusts
Category 1 (Category 1 subsumes the requirements for Category 2 and Category 3)	Very high	Two independent means of protection are incorporated. The design ensures that the maximum surface temperature will not be exceeded even in the most unfavourable circumstances	An explosive atmosphere of gas/vapour/haze/dust is continuously present or present for long periods (more than 1000 hours per year)	Zone 0 (gas) and Zone 20 (dust)
Category 2 (Category 2 subsumes the requirements for Category 3)	High	The design and construction prevents ignition sources even in the event of frequently occurring disturbances and equipment operating faults. The design and construction ensures that the maximum surface temperature will not be exceeded even in the case of abnormal situations anticipated by the manufacturer	An explosive atmosphere of gas/vapour/mist/dust is likely to be present (between 10 and 1000 hours per year)	Zone 1 (gas) and Zone 21 (dust)
Category 3	Normal	The equipment must be designed and constructed so as to prevent foreseeable ignition sources likely to exist during normal operation	An area where an explosive atmosphere of gas/vapour/mist/dust is unlikely to occur and if it did it would be for a short time only (less than 10 hours per year)	Zone 2 (gas) and Zone 22 (dust)

Table 4.17 Types of protection for lighting systems for use in hazardous areas, their symbols, the typical categories to which they are suited and their method of protection

Type of protection	Symbol	Typical ATEX category suited to type of protection	Method of protection
Increased safety	e	M2 and 2	The equipment is designed to
Non-sparking	nA	3	ensure that no arcs, sparks or hot surfaces are produced
Flameproof	d	M2 and 2	The equipment is designed to
Enclosed break	nC	3	contain the explosion and
Quartz/sand filled	q	2	prevent flame propagation
Intrinsic safety	ia	MI and I	The equipment is designed to
Intrinsic safety	ib	M2 and 2	limit the energy of any spark
Energy limitation	nL	3	and the surface temperature of the equipment
Pressurised	p	2	The equipment using these
Restricted breathing	nR	3	types of protection is designed
Simple pressurisation	nP	3	to ensure that the flammable
Encapsulation	m	2	gas is kept out of the
Encapsulation	ma	I	equipment
Oil immersion	o		

typical categories to which they are suited and their method of protection are given in Table 4.17.

4.7.5 Selection of equipment for use in hazardous areas

4.7.5.1 Selection considerations

Selection of lighting equipment for use in hazardous areas must include consideration of: the zone and category that are applicable; the relevant gas group; and, the T rating. Another important consideration in the selection of lighting equipment is the nature of the environmental conditions in which the equipment will operate. These may be hostile as well as being hazardous, and involve the equipment experiencing extreme high or low ambient temperatures. Manufacturers will provide information on the ambient temperature range, T_{amb}, in which their product will operate safely and satisfactorily. A typical T_{amb} might involve the range $-20\,°C$ to $+40\,°C$. In selecting equipment it is important to know that it is certified for the operating conditions of the environment.

4.7.5.2 Ingress and mechanical protection

The equipment may also be subjected to liquid spray (offshore lighting) or dust for lengthy periods. Maintenance of the equipment might require the removal of dirt deposits through washing down using high-pressure water jets. If the equipment is to be exposed to the vagaries of an outdoor environment, it is important to know the extent to which it protects against the ingress of solids and liquids. In hazardous areas during normal usage the equipment may also be subject to impact. It is also important to know the extent of mechanical protection that is offered by the equipment. Details of how protection against impact and ingress of material are measured and indexed are given in Section A.2.

4.7.5.3 Product marking and identification

Electrical lighting products must be clearly marked to indicate their suitability for operation in hazardous areas. Those responsible for installing and maintaining lighting equipment need to be fully aware of the significance of the markings so as to ensure the equipment is entirely suited to its purpose and area of operation. The markings on the equipment will show the ATEX coding and should also give coded information on the harmonised standard against which product has been certificated. A description of the meaning of the ATEX codes used in a typical equipment marking are shown in Figure 4.21.

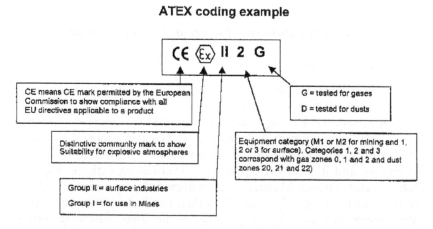

ATEX coding example

Figure 4.21 Description of the meaning of the ATEX codes used in a typical equipment marking. (Reproduced by permission of TRL.)

4.7.6 Recommendations

The maintained illuminance levels tabulated in Table 4.18 are recommended as a basis for the design of new installations, unless higher levels are required by Regulations in the country of installation.

4.8 Sport

4.8.1 Introduction

The increasing amount of leisure time available to the public, coupled with a wider interest in health and fitness, has led to an expanded demand for sports facilities. The demand is being met by the provision of well-equipped facilities in cities and towns across the country. For outdoor sports the economic need to extend the hours of use necessitates the inclusion of specialist lighting schemes geared to the sport concerned, and the level to which it is played.

For multi-sport use the lighting needs to be appropriate to all the various activities. In major stadia, the lighting needs of television cameras must also be met, and in these facilities the maintenance of safety levels for large numbers of spectators must be considered.

Concerns about the environment have led to refinements so that the lighting for a sports facility does not generate discomfort or annoyance to local residents. Light spillage beyond the facility itself, and glare from luminaires, are common sources of concern, and the prevention of light pollution is increasingly a requirement of planning authorities. Section 2.3 gives guidance on the avoidance of light pollution.

Lighting a playing surface to a general level is no longer enough if all these demands are to be satisfied, and sports lighting, in seeking to meet the many different needs of activities and participants, needs careful design.

4.8.2 Standards and codes of practice

Many sporting bodies issue guidance on lighting levels for their particular activity. The lighting engineer should consult with the relevant sports authority to determine their recommendations. Some information sources are given in the Bibliography.

It should be borne in mind that many of the lighting codes specify only lighting levels on the playing surface, and therefore offer little more than a numeric target figure to be achieved. Most codes also specify a uniformity standard but achieving a lighting level on the pitch alone addresses only a part of the problem. Wider aspects of good lighting practice demand a more detailed approach than just meeting a lighting and uniformity standard on the horizontal surface.

Table 4.18 Recommended lighting levels for hazardous areas

Area	Maintained average illuminance (in lux) *not less than*			Working plane height (m)	Minimum maintenance factor
	Normal	Emergency	Escape		
Oil rigs					
Exterior walk-ways/stairs/access ways	50	20	2	0.0	0.80
Under and around platform	50	25	0	sea	0.75
Lifeboat/life raft stations	150	100	5[a]	0.0	0.75
Weather deck	50	20	0	0.0	0.75
Laydown areas	100	20	2	0.0	0.75
Muster point	150	50	10	0.0	0.85
Auxiliary rooms: Outside, near entrance	150	20	2	0.85	0.85
General industrial including oil rigs					
Process plant main pumps	200[b]	25	2	0.0	0.85
Process plant main compressors	200[b]	50	2	0.0	0.85
Process plant general	150[b]	50[b]	2	0.0	0.80
Workshops: Outdoor storage area	50	25	2	0.0	0.85
Warehouse and stores: Outdoor storage area	50	25	2	0.0	0.85
Petrol stations					
Approach, apron and general	50		2[c]	0.0	
Access to forecourt, forecourt and pump areas	300		2[c]	0.0	

Notes
a Illuminance levels to be measured in the vertical plane.
b Where overhead travelling cranes are installed, floodlights should be fitted under the crane beams to provide an illuminance level of 400 lux, for better illumination during maintenance.
c Only required if remote from public road with fixed road lighting.

4.8.3   The visual task

In planning a scheme, it must first be determined what the lighting will be required to illuminate.

Primarily the lighting is needed for the players so the sport can take place, but different levels of play, the type of facility and the needs of others at the game must be considered. The key factors are shown in Table 4.19.

Therefore a game played at amateur level, on an open pitch, will have limited factors to consider. These would be the needs of the players and officials, and also environmental factors, mainly the prevention of light pollution, that is upward light, spill light, and intrusive glare to neighbouring premises.

At the other end of the scale, lighting for a sports stadium must additionally meet the needs of many spectators, TV cameras, and also provide a safe environment for the period the site is populated. This will include the provision of emergency lighting to operate in times of power failure.

4.8.4   Principles

The following factors will determine the lighting standards required for each particular sport.

4.8.4.1   Lighting the playing surface

Lighting of any large horizontal area requires the projection of light from a source mounted in an elevated position. Sports lighting is characterised by the light being projected from positions which themselves are outside the playing area. This calls for precision in the distribution of the light

Table 4.19 Prime lighting considerations for different facilities

	Level of play		
	Practice/amateur	County/club	Professional
Type of pitch	(open)	(open/semi enclosed)	(stadium)
Factors			
Players	*	*	*
Referee/linesmen/judges	*	*	*
Spectators		*	*
TV cameras			*
Safety and security[a]		*	*
Environmental matters[b]	*	*	*

Notes
a This is mainly concerned with emergency evacuation procedures.
b Usually to do with light pollution, and glare to adjacent premises.

output from a luminaire if this light is to be accurately distributed across the playing area, and if the lighting is to be restricted to that area. For generic types of luminaires having specific applications, see Section 3.4.2.

Lighting levels for any sport are primarily determined by the size and speed of movement of the ball. Lighting for tennis, for instance, will have more stringent requirements than for, say, soccer, where the ball is larger and slower moving. For players, rapid and accurate sighting of the ball is the critical factor, and must be seen as the prime reason for sports lighting. The lighting requirement is also influenced by the level of play; higher levels requiring a higher degree of visual accuracy.

Similarly, uniformity is a requirement determined by the type of sport. With a more demanding visual task, i.e. a small and/or faster moving ball, uniformity requirements are increased.

The control of glare is a requirement from both the spectators as well as the player's viewpoint, and is achieved by carefully controlling the distribution of light output from the luminaires. For higher levels of play, glare should be lower.

Finally, good colour rendering is important. Lamps with an R_a up to 20 may be used for lower levels of play. However, an R_a greater than 60 is recommended.

4.8.4.2   Horizontal and vertical illuminance

The above refers only to lighting performance as measured on the horizontal plane. Where the ball commonly has a high trajectory, for instance in rugby or tennis, it is vital that players can follow the ball, and so lighting in the vertical plane is also important.

Equally, where TV cameras are in use, they will 'see' the vertical surface of players as well as the horizontal pitch surface. For this reason, there would normally be a minimum vertical illuminance specified, as well as a specified ratio between the horizontal and vertical illuminances within the project.

Most codes of practice for sports played at amateur level do not specify any levels of vertical illuminance. However, if the designer restricts his/her calculations to the provision of light to horizontal surfaces only, then the effectiveness of the scheme may be less than ideal. BS EN 12193 specifies that the vertical illuminance should be at least 30 per cent of the horizontal illuminance.

4.8.4.3   Distance from playing surface

The inverse square law decrees that the illumination on a subject is a function of the intensity of the light projected towards it, and the distance of that projection. For a given intensity, doubling the distance results in one quarter of the illumination. With the typical large distances between the luminaire and the area to be lit involved in sports lighting, there is a need for very intense sources.

For example, a luminous intensity of 100 000 cd, directed at a surface 50 m away, will produce an illuminance

$$E = \frac{I}{d^2} = \frac{100\,000}{50 \times 50} = 40 \text{ lux.}$$

Therefore luminaires will have a declining effect with increased distance. Consequently, mounting positions as close to the playing surface as are allowed by safety considerations should be sought.

4.8.4.4 Angle of incidence on playing surface

The cosine effect determines that light incident on a surface will produce less illumination if the light arrives at any angle other than perpendicular to the surface. So for sports lighting, where a luminaire delivers light at a high angle of incidence, the amount of luminous intensity required to produce adequate levels of illumination is increased. Using the example above, if the light arrives at an angle 30° away from perpendicular, the full calculation becomes

$$E = \frac{I}{d^2} \times \cos 30° = 40 \times 0.866 = 34 \text{ lux.}$$

Therefore luminaires should be at the highest reasonable mounting height to reduce the angle of incidence.

4.8.4.5 Light pollution

This topic is dealt with fully in Section 2.3, but sports lighting demands special attention because of the high intensity, high output lighting equipment, mounted at extended heights, which is in use.

Because of the relatively high levels of lighting used on a playing surface, some sky glow caused by reflection is unavoidable. However, it is desirable to select luminaires which produce no upward light. At the same time, lighting the playing surface only to the standard level, and no more, limits the light available for reflection. In some cases it has been the practice to light to an average level above that required, in order to achieve the uniformity standard called for. With the correct choice of lighting equipment this should rarely be necessary.

Light trespass and intrusive light can both be controlled by proper design of a scheme, and accurate aiming of the lighting equipment. Sometimes, auxiliary shields may be required as fitments to a luminaire so that fine-tuning can be achieved during the commissioning phase of a project. It is unacceptable if an observer outside the lit area can see the light source within the luminaire.

Light pollution, and the measures taken to limit the effects, will form part of the assessment carried out by planning authorities when a scheme is presented for approval, and data generated from the scheme will be required to satisfy them that the subject has been adequately addressed. Most projects will need individual detailed design work since light pollution will be closely related to site position and local circumstances.

4.8.5   Design considerations

Common design features are as follows:

4.8.5.1   Corner towers vs side lighting

For non-stadia facilities, the lighting demands for most sports are easily achieved using positions at or near the corners of the area. Tennis, for example, benefits from corner mounting in that the rear surface of the ball is clearly visible as it travels away from a player. A player's ability to track the shot that has just been made is vital to the game. It should be noted that many planning authorities are applying a restriction of 12 m maximum tower height for supporting structures, in line with their concern for protecting skyline views. This can cause some problems when the designer wants to use a flat-glass type luminaire and still project adequate quantities of light across a wide area. To do this, a luminaire with a high-peak beam angle is required, probably around 65–70°. Planning authorities and engineers should consider the balance between the daytime intrusion from increased mounting height and the lowering of light pollution and spillage resulting from higher mounting heights.

Higher mounting positions will usually mean higher capital cost in the towers. Mounting on the roof of a stand limits the height, but this may be increased by the use of a superstructure on the top of the stand itself.

If the peak beam angle is insufficient, then the light will not reach far enough across the playing surface. Tilting a flat-glass luminaire to increase the range may defeat the object of using a flat-glass luminaire by potentially allowing upward light.

For stadia facilities the position is different. Early forms of pitch lighting relied very heavily on mounting luminaires on towers positioned at the four corners of the playing area. This format produced the familiar 'four shadows' scene. There were some significant problems, however, with delivering sufficient light along touch lines. The towers could also restrict spectator views. The achievement of uniformity was usually addressed by the use of very high towers, so that the light was incident on the pitch at a more advantageous angle (nearer the vertical).

Lighting a pitch from side positions only, using wide flat beam luminaires, has become more popular, as this produces better results for the players, and less obstruction for spectators. The luminaire required will have a beam which is wide in the horizontal axis, and strongly asymmetric in the vertical axis. This type has become more popular with the development of linear light sources. With covered stands, the use of roof-mounted luminaires has become a standard feature of stadia. Side lighting also achieves more easily the lighting of vertical surfaces, which is such a vital component where television cameras are used.

Side lighting is therefore now the most common format both for county/club levels as well as major stadia (although in the latter case the lighting may be roof mounted on all four sides).

4.8.5.2 Luminaire types

For sports lighting, there are three generic types of luminaire, and these are described below:

- Symmetrical projector (see Figure 3.19) – Characterised by having a circular reflector system and a lamp with a compact discharge tube. This type produces a narrow beam of conical shape, suitable for long projections, and was popular in applications using corner tower format.
- Asymmetric projector (see Figure 3.21b) – Characterised by having a rectangular reflector system, and a linear discharge-type lamp. The resultant beam is wide, and strongly asymmetric in the vertical plane. Internal baffles are required to limit light above the beam peak intensity. This is the most common type in side-lit applications, particularly where enclosed stands are used, since the aiming capacity of the projector can be used to limit glare for spectators on the opposite side of the ground.
- Flat-glass projector (see Figure 3.21a) – This type is designed to operate with the front glass parallel to the ground, and can be useful in limiting upward light, if they are used in such a position. Flat-glass types tend to be less efficient in lighting an area, but where light pollution is a critical factor then they are recommended. On 'open' pitches where fixed stands are not present and there is therefore no natural barrier to light spillage beyond the pitch, then the flat-glass type can be useful in restricting the spread of light beyond the pitch surface.

4.8.5.3 Operating temperatures

The lamp and the control gear generate heat. The rate of heat dissipation will depend on the operating ambient temperature, as well as other factors such as the materials and construction of the luminaire. In outdoor facilities the generation of heat will rarely be a problem as ambient temperatures are lower than for internal applications, and the unrestricted surrounding air space will aid the dissipation of heat.

4.8.5.4 Emergency lighting

It is a requirement of public facilities that safe evacuation can be effected in times of emergency, and where spectators are present within a sports facility then this requirement must be met. Emergency lighting may be in the form of self-contained units with integral battery-powered light sources, which are automatically operated during mains failure, or it may take the form of auxiliary lighting powered from an independent electrical supply (e.g. a generator set). In a stadium, emergency lighting within the stands and throughout the escape routes will be required.

For typical local club facilities, where seating is not enclosed, emergency lighting may not be a requirement. Local Fire Safety Officers will be able to offer advice on the possible need for emergency lighting.

4.8.5.5 Safe stopping

There is also the need to consider the safety of players in the event of mains power failure. Some sports will expose the players at considerable personal risk if they are suddenly plunged into darkness. For this reason, many sports have a 'safe stopping' requirement which specifies the length of time that at least a given percentage of normal illumination must be maintained if the lighting equipment is suddenly extinguished. For cycle racing, for instance, it is required that at least 10 per cent of the normal lighting level must be maintained for at least 10 s after mains power failure. This is deemed sufficient to enable participant's time to stop safely.

4.8.5.6 Control gear positioning

All discharge lamps require control gear for both starting and operating (see Section 3.3). Because of the need to place as little weight and windage load as possible at the top of a column, it is normal to retain the ignitor with the luminaire (it needs to be relatively close to the lamp to avoid voltage drop), and to position the rest of the control gear at the base of the column, or in an adjacent cabinet. This is particularly true with 1 and 2 kW luminaires. For luminaires of lower wattage, it is possible to use types where the gear is integral, since the physical size and weight of the gear do not pose a problem in mechanical loading of the column.

4.8.5.7 Hot re-strike luminaires

High intensity discharge lamps will extinguish from even a momentary loss of power supply, and will then need to cool down before the starting cycle can commence. During this time there is initially no light output, and full output will be some minutes after the commencement of the starting cycle.

Where this is likely to cause a problem, e.g., at professional standard games, and/or where television broadcasting is involved, then provision must be made to give some alternative lighting provision.

This can be achieved by the use of standby generators, but increasingly 'hot re-strike' types of luminaires and control gear are being used. These types have the facility to immediately re-start the lamp after a momentary power loss and recovery.

4.8.5.8 Recommendations

The following recommendations are made for the lighting of popular exterior sports. The selection of lighting class for sports is dependent upon the level of competition and should be determined from Table 4.21. The lighting standards for many sports will specify a principal area, which is the actual playing surface, and a total area, which also embraces an additional safety area outside the playing surface. Unless otherwise specified, the extra area should be lit to 75 per cent of the principal area illuminance.

Template schemes for popular sports are shown in Section A.3.2.

4.9 High mast lighting

4.9.1 Introduction

High mast lighting covers the use of tall columns (high masts) fitted with clusters of luminaires to illuminate large areas.

At present BS EN and CIE specifications do not provide a basis for lighting design when luminaires are placed irregularly, in clusters and at ultra-wide spacings. Conventional luminance-based specifications do not apply to high masts as the complexity of the situations where they are used, their irregular positioning and their extreme heights and spacings fall outside the normal limits in those specifications.

In order that a scientifically justifiable specification can be used to judge the lighting schemes offered by different designers on a common basis, it has

Table 4.20 Lighting recommendations for sport

Sport and area to be illuminated	Level of play	Maintained average illuminance (in lux) not less than		Uniformity ratio $E_{h\,min}/E_{h\,ave}$ not less than	Glare rating GR not greater than	Colour rendering index R_a not less than
		Horizontal value (E_h)	Vertical value (E_v)			
Archery						
Shooting lane		200		0.5		60
Target			750	0.8		60
Artificial ski slope		50		0.3	55	20
Athletics	I	500		0.7	50	60
	II	200		0.5	55	20
	III	100		0.5	55	20
Cricket (Square/infield)	I	750		0.7	50	60
	II	500		0.7	50	60
	III	300		0.5	55	20
Cricket (field/outfield)	I	500		0.5	50	60
	II	300		0.5	50	60
	III	200		0.3	55	20
Cycle racing	I	500		0.7	50	60
	II	300		0.7	50	60
	III	100		0.5	55	20

Equestrian	I	500	0.7	50	60
	II	200	0.5	55	20
	III	100	0.5	55	20
Golf driving ranges					
General		100	0.8		20
Distance marker		50			20
Hockey	I	500	0.7	50	60
	II	200	0.7	50	20
	III	200	0.7	55	20
Kabaddi	I	500	0.7	50	60
	II	200	0.6	50	60
	III	75	0.5	55	20
Netball	I	500	0.7	50	60
	II	200	0.6	50	60
	III	75	0.5	55	20
Pétanque and Boules	I	200	0.7	50	60
	II	100	0.7	50	20
	III	50	0.5	55	20
Rugby	I	500	0.7	50	60
	II	200	0.6	50	60
	III	75	0.5	55	20

Table 4.20 (Continued)

Sport and area to be illuminated	Level of play	Maintained average illuminance (in lux) not less than		Uniformity ratio $E_{h\ min}/E_{h\ ave}$ not less than	Glare rating GR not greater than	Colour rendering index R_a not less than
		Horizontal value (E_h)	Vertical value (E_v)			
Skateboard parks						
	II	150		0.5		60
	III	50		0.4		20
Soccer						
	I	500		0.7	50	60
	II	200		0.6	50	60
	III	75		0.5	55	20
Tennis						
	I	500		0.7	50	60
	II	300		0.7	50	60
	III	200		0.6	55	20

Source: Reproduced by permission of BSI.

Table 4.21 Lighting class for level of competition

Level of competition	Lighting class		
	I	II	III
International and National	*		
Regional	*	*	
Local	*	*	*
Training		*	*
Recreational			*

Source: Reproduced by permission of BSI.

been felt necessary to produce guidance notes for where high mast lighting is used.

This section covers the design of high mast lighting schemes, not the engineering of high masts, and should be read in conjunction with the ILE Technical Report No. 7, High Mast for Lighting and CCTV which covers the design, manufacture, assembly, erection, painting, testing and maintenance of high masts.

Guidance is given on the general principles of high mast lighting with mounting height between 18 and 50 m, and recommendations provided for the lighting of large areas.

The areas considered are

- Roads – including multi-level junctions
- Ports
- Airports
- Car parks.

4.9.2 Use of high mast lighting

High mast lighting is used in many situations:

- To illuminate wide areas where columns or other supports for luminaires cannot be positioned (e.g. an airport apron)
- To illuminate a large area, with the minimum number of obstructions (e.g. a container port)
- To illuminate complex multi-level road junctions without a forest of columns (e.g. a motorway intersection) and
- To illuminate sporting facilities for players, spectators and television (e.g. an athletics arena).

High mast lighting should provide the specified illumination level with lower glare and a higher degree of uniformity when compared to conventional lower height solutions.

4.9.3 General design recommendations

4.9.3.1 General

The main purposes of lighting for the areas covered by this section are to enable vehicles and pedestrians to orientate themselves and undertake actions within the illuminated area, to discourage and enable detection of criminal activity and to provide a safe working environment.

4.9.3.2 Design principles

Luminance design is not generally compatible with high mast lighting. High mast lighting is generally used in areas where the specification of a single viewer looking in one particular direction is not relevant due to the large variety of viewing directions and positions.

The area illuminated by each high mast is large and it is poor practice to use a small number of luminaires of the maximum light output to provide the specified illumination level. If in such an instance the lamp in a critical luminaire should fail, an important part of the defined area may be in darkness. If possible, all parts of the defined area should receive illumination from more than a single luminaire.

4.9.3.3 The specified task

The illumination levels should be designed with respect to the action normally undertaken in the defined area (the specified task). Additional illumination may be required if work more detailed than the specified task is to be done, and the relevant recommendations for that task should be provided on a temporary basis.

4.9.3.4 Luminaires

All luminaires used for high mast lighting should be of the full cut-off type emitting no light above the horizontal plane. The reason for this is that high masts will be visible from afar and the light emitted at angles near to or above the horizontal will be more visible at night than luminaires mounted at a lower height. Sports lighting may require the use of floodlights, not of a full cut-off design, and as an arena is only in use for a limited number of hours, this may be justified.

Individual luminaires may provide a symmetrical or an asymmetrical light distribution. They should be grouped in such a manner that the light distribution from the high mast matches the area to be illuminated as closely as possible.

4.9.3.5 Light sources

Light sources used in high mast lighting need to be of a high intensity with a long life. For this reason HID lamps of the high pressure sodium or metal halide type are used. High pressure sodium lamps will have a greater luminous efficiency and a longer life, which makes them the preferred light source. Where the task requires a light source with a better colour rendering, metal halide lamps can be used. However, their shorter lamp life and greater lamp lumen depreciation must be considered both at the design stage and when planning future maintenance. It may be necessary to use supplementary tungsten halogen or instant restrike discharge lamps in situations where the delay in hot restrike of standard discharge lamps after a power failure is unacceptable for security or safety reasons.

4.9.3.6 Mast height

The mounting height on a high mast design will depend on the area to be lit. Factors that will affect the decision include

- High obstructions within the area which will shadow part of the area
- Fixed positions, which will require masts of a certain height to illuminate the area between them and
- Limitations on height imposed by third parties, which may require a lower than ideal height to be used.

4.9.3.7 Mast position

The position of the masts will be influenced by a number of factors and the ideal photometric arrangement is rarely achieved. The following list indicates some factors that will influence mast positioning; however, it is not and cannot be comprehensive and rarely encountered factors or ones unique to a site must be considered:

- Ground obstructions (e.g. road or car park layout, landscaping, buildings, rail layout)
- Ground conditions (e.g. sub-surface obstructions, poor soil, or steeply sloping ground)
- Overhead obstructions (e.g. power transmission cables)
- Access for maintenance

- Safety (a potential traffic hazard)
- Daytime visual appearance (e.g. not to obscure a view to or from a building, to fit an architects overall plan) and
- Shadowing.

4.9.4   Practice in relation to type of location

4.9.4.1   Roads

Luminance design is not generally compatible with high mast lighting. High mast lighting is generally used in areas such as multi-level junctions and roundabouts, where the specification of a single viewer looking in one particular direction is not relevant. The infinite variety of viewing directions and positions possible makes luminance design irrelevant for all but a very few schemes.

Although high masts are not often used for lighting of individual roads, for long straight sections of road it is possible to specify luminance. The lack of regular spacings and irregular mast positioning mean that a section of road over which the calculations are to be performed must be specified, together with the observer position for any luminance calculations. The calculation methodology performed under EN 13201 does not apply and recommended levels are given in Table 4.22. The CE lighting classes from EN 13201 should be used where relevant, especially for conflict areas such as roundabouts, see Table 4.28.

Where individual luminaires are mounted at regular spacings in a standard arrangement despite being higher than what the standard allows, the luminance values specified in EN 13201 can be used.

4.9.4.2   Ports and harbours

The lighting of ports and harbours is covered by the Docks Regulations (Table 4.23 – for lighting recommendations).

The height of the mast to be used will depend on the spacing of masts and the height to which the containers or bulk cargo are stacked. The problem of shadowing and small number of mast positions, to enable maximum

Table 4.22 Lighting recommendations for roads illuminated using high masts

Area	Maintained minimum point illuminance (in lux) not less than	Uniformity ratio not less than
Motorways	20	0.5
Multi-level junctions	20	0.4

Table 4.23 Lighting recommendations for ports and harbours

Area	Maintained minimum point illuminance (in lux) not less than	Uniformity ratio not less than
Ferry vehicle parking (short term)	30	0.3
Ferry vehicle parking (long term)	15	0.3
General lighting outside working hours	10	0.25
Quays and jetties	50	0.15
Storage areas	10	0.40
Loading, unloading and handling general cargo	20	0.25
Passenger movement, gangplanks and ladders	50	0.4
Working areas on quays	20	0.25
Movement of vehicles	20	0.4
Passage of dock workers	10	0.15
Container terminals	20	0.25
Gauge reading, handling pipe couplings	50	0.4

flexibility and manoeuvrability of the heavy cranes and large lorries, often causes very high masts to be used, e.g. 45 m height.

As ports are visible from great distances all luminaires used for high mast lighting should be of the full cut-off type, emitting no light above the horizontal plane.

4.9.4.3   Airports

(The term 'Aerodrome' is used in the Civil Aviation Act 1982 to describe an airport, airfield or airstrip.)

The specification for lighting on and in the vicinity of airports in the UK is covered by CAP168 'Licensing of aerodromes' published by the Civil Aviation Authority. ICAO, Annex 14 is the international standard (Table 4.24 – for lighting recommendations).

When providing lighting on or near airports it is critical that no light which is liable to endanger aircraft taking off or landing or which can

Table 4.24 Lighting recommendations for airports

Area	Maintained average illuminance (in lux) not less than	Maintained minimum point illuminance (in lux) not less than
Aprons	20	5
Service areas	50	20
Car parks:		
short term	30	10
Long term	15	5
Passenger pick up and put down areas	50	20
External pedestrian areas – landside	30	15

be mistaken for an aeronautical light is displayed. This may include lights where:

1. The intensity causes glare in the direction of an approaching aircraft.
2. The colour (e.g. advertising signs) may cause the light to be mistaken for an aeronautical light.
3. When viewed from the air the lights make a pattern similar to an approach or runway lighting pattern (e.g. a row of street lights).
4. The overall level of illumination detracts from the effectiveness of any visual aids provided by the aerodrome for use by aircraft, particularly in poor visibility.

It is particularly important that lighting is provided with a minimum of glare to pilots of aircraft in flight and on the ground, aerodrome and apron controllers, and personnel on the apron. Therefore full cut-off luminaires should be used.

Care should be taken to ensure that no structure carrying lighting equipment impinges on the height restrictions imposed on and around the airport. This area is referred to as a safeguarded area and is normally circular, centred on the mid point of the runway(s) but may be race-track shaped if the runway(s) are longer than a certain length. The local planning authority and the airport operator should be consulted on the safeguarding area and restrictions.

All developments in the vicinity of a military site must be referred to the ministry for Defence (Safeguarding, Defence Estates).

4.9.4.4   Car parks

For car parks refer to Section 4.13.

4.10 Festival lighting

4.10.1 Introduction

This section outlines the requirements and considerations to be taken into account when proposing to erect festival lighting in public areas. Christmas is the main Christian festival in the Northern Hemisphere that extensively uses lighting. However, there are many other religious and non-religious festivals that make extensive use of lighting. These together with lighting along sea fronts etc. are aimed at making a location special to the occasion or promoting and encouraging public attendance to an area or attraction.

The fundamental aim of any illuminated festival decoration is to provide a brighter and more interesting environment at night, creating an atmosphere that makes the location special for the occasion. Unfortunately, an enriched night-time environment is often provided at the detriment to the daytime appearance. When planning festival lighting, the daytime appearance of the decoration should be given equal importance and both aspects should be carefully considered at the design stage.

4.10.2 Various forms of festival decorations

Festival decorations can take various forms, such as:

- Simple festoon with multi-coloured lamps
- Decorations mounted on poles or lighting columns
- Area spanning decorations mounted on catenary wire attached either to buildings or poles
- Ground mounted features
- Christmas trees
- Additional effect lighting and
- Laser displays.

4.10.2.1 Decorations and festoons

These decorations will generally consist of moulded internally illuminated features, decorations made from tinsel and festoon lighting wrapped around a metal/plastic frame, strings of lamps mounted on festoon cable or ground mounted tableaux with external lighting.

The precise type of decoration or festoon should be chosen to suit the desired effect and consideration may be given to the use of lit elements of decorations that are constructed from rope light.

Where rope light is not used, the system should have replaceable lamps, with individual lamp failure bypass circuitry to ensure operation following a lamp failure. In addition, decorations should be constructed with a combination of series and parallel circuits.

4.10.2.2 Effect lighting

Additional effect lighting to, for example, significant buildings, statues, clocks, etc. can significantly add to the atmosphere at festivals. The principles of effect lighting are detailed in Section 4.5.

4.10.2.3 Laser displays

Other types of light source such as xenon arc lamps, lasers and other finepoint light sources can be considered for specialist moving light effects. These sources, which in some cases can explode if handled incorrectly, should be regarded as potentially hazardous and only installed and operated by competent and trained operatives.

Consideration must be given to the optical output of lasers and the client for the display must ensure that such displays are installed, programmed and operated correctly to ensure the safety of not only the public from normal viewing angles but occupiers of adjacent properties, who may view the display from an unusual angle.

The following aspects must all be carefully considered and complied with:

- Planning
- Safety policy
- Aviation
- Laser system
- Installation and operation
- Maintenance and testing
- Statutory and other requirements.

Greater detail with regard to the use of festival lighting and to the use of laser displays can be obtained from ILE publication 'Lasers, festival and entertainment lighting code'.

4.10.3 Health and Safety

Whoever is the client for the installation and operation of festival decorations, be it the Council, Traders' Association, Chamber of Commerce or other, assumes the main responsibility for health and safety. The client must ensure that all decorations are correctly supported and electrically safe and do not present a hazard to the public. To achieve this they must obtain detailed information regarding specifications, electrical loading, weight and wind loadings.

They must ensure that the routine inspections and electrical inspection and tests are carried out as required (see Section A.3.3).

Where electrical connections are to be made to existing lighting circuits, it must be the responsibility of the festival lighting contractor to procure the appropriate permits. Structural considerations for the erection of decorations on columns or poles are detailed in Section A.3.4.

4.10.4 Approval for erection of decorations

Where decorations are to be fitted to private property the appropriate consents, approvals and wayleaves must first be obtained. For decorations fitted to local authority highway furniture approval must first be obtained from the highway authority. Mountings should be structurally tested in advance of each erection process (i.e. usually annually) and should include any small-scale stonework to which mountings are attached, e.g. brick or sandstone slabs.

Highway authorities may require a certificate of insurance obtained by the installer indemnifying against any third-party claims relating to the installation, removal or operation of the decorations and any damage to Council property.

4.10.5 Highway safety

Decorations that contain flashing red, amber or green lamps should not be installed within the visual area of traffic lights or signal-controlled pedestrian crossings.

All decorations, festoons, etc. mounted above areas accessible to vehicles must have a minimum clearance above ground level, at their lowest point, of 5.8 m.

Decorations must be electrically inspected and tested in advance of each erection process (i.e. usually annually) and before switch-on.

4.10.6 Switch-on

It is possible by use of connecting cables or radio-activated contactors to energise festival decorations over an area at least as far as the eye can see in an urban environment. If a formal switch-on ceremony is taking place this equipment should be considered to achieve maximum effect. Whether or not there is to be a formal switch-on ceremony, a trial switch-on should be carried out at least four working days in advance of the planned switch-on date in order that any defective equipment can be repaired. Where there is a formal ceremony, this also allows operatives to be familiar with procedures and modifications to be carried out if necessary.

4.10.7 Maintenance

Maintenance of the installation is an essential part of the process as the effect can be spoilt by poor standards, and safety can be compromised. A visual inspection should be carried out at least twice a week and any circuit failures resolved as soon as possible. The recommended inspection frequency is every second night. Acts of vandalism should receive prompt attention.

4.10.8 Removal

All items of equipment should be labelled according to a pre-prepared schedule as they are being removed to allow easy identification during the next erection. Damaged equipment should be identified for repair or renewal. Any cable entry holes or sockets in lighting columns, posts or other apparatus should be appropriately sealed to prevent water ingress.

4.11   Traffic routes

4.11.1   General principles

The lighting of traffic routes and motorways differs from other outdoor applications in that levels are specified in luminance rather than illuminance. This is a more complete criterion than illuminance and it is possible to use it because a reasonable assumption can be made about the positions of the eyes of the principal observers, i.e. at approximately 1.5 m high in a traffic lane.

The principle is to provide a reasonably uniformly bright road surface against which the driver of a vehicle will see small dark hazards, between 60 and 160 m ahead, in silhouette, with no assistance from the vehicles dipped headlights.

The driver of a moving vehicle has to absorb sufficient visual information from the continually changing view in front of him so that he can proceed safely at a reasonable speed, see his route ahead, respond to signs and manoeuvre in good time.

As only the small central part of the field of vision of a driver will be in sharp focus, most information will be received peripherally and, therefore, not in detail. The significant object in a driver's field of vision should cause the driver to focus their attention on it but for a driver to detect the presence of an object the object must have sufficient contrast against its background. This is true both by day and night, but at night the driver's ability to perceive contrast is considerably reduced at lower-lighting levels. Road lighting should therefore aim to make the general scene as bright as possible to maximise the contrast between objects and their background.

Whereas in most lighting the aim is to light objects of interest rather than their background, in road lighting the converse is true. In this way a relatively small amount of light can be used to maximum effect by lighting the road surface and the immediate surrounds against which objects will generally appear in silhouette. The success of this method of lighting depends on designing the distribution of the light from the luminaires to take advantage of the reflection properties of the road surface. Although luminance design is more precise than illuminance, inaccuracies remain. Whereas the quantity and direction of the light omitted by the luminaires is known, the reflective properties of the road surface can only be determined to a much lower level of accuracy.

4.11.2   Road surfaces

The reflective properties of the road surfaces are represented on graphs according to the specular factor S_1 and the luminance coefficient Q_0 of the different surfaces. The CIE have specified four classes, which are represented by the standard surface types called R1 to R4 respectively based on these parameters. In a statistical exercise it has been shown[5] that a system using this limited number of road-surface classes based on one specular factor gives adequate accuracy for practical purposes when compared to a system using two specular factors and a larger number of classes. It should be noted, however, that such accuracy does require a surface of the correct class to be selected. The surfaces can be plotted according to the classification system as shown in Figure 4.22 to help illustrate their reflective properties with respect to each other.

Those surfaces with higher Q_0 values reflect more light and are towards the top of the graph. Those with higher S_1 values are more specular, that is more like a mirror, and towards the right of the graph. In some places where measurements of the local road surface reflection characteristics have been carried out, variations on the standard surfaces are used, e.g. R1 but with a Q_0 value of 0.08 rather than 0.1. If the wrong road surface is used in luminance calculations and the results found are then implemented in a project, either the lighting will be sub-standard or there will be waste due to over provision.

4.11.3   Appearance

The appearance of the lit roadway results from the superimposing of the bright patches formed by the luminaires. The design of the lighting installation involves positioning the luminaires so that the bright patches join to give the road surface an average luminance and overall and longitudinal

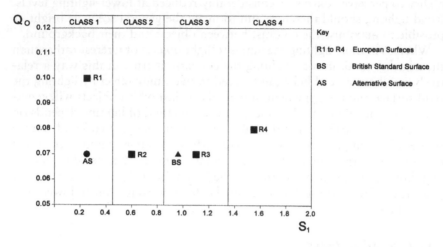

Figure 4.22 Relationship between S_1 and Q_0 for a variety of road surfaces.

uniformities which are satisfactory for the use of the road. Since the aim of the lighting is generally to reveal objects in silhouette, the luminance of vertical and nearly vertical surfaces should be as low as possible. This is achieved at most positions on the road as a result of the interrelation of the spacing of the luminaires, their mounting height and their light distribution. A light-coloured object may appear in reversed silhouette and there will be a certain reflectance of the object which will give it the same luminance as that of the road forming the background, making it disappear. In practice, this is unlikely to occur – first because objects are viewed against a considerable stretch of road or road surrounds, which will show some unevenness of luminance; and second because objects are rarely of uniform reflectance. In addition, the situation is dynamic; as the driver changes his position the angles of view and reflected light also change. It is also important to light the surrounds of the road both to assist pedestrians and allow the driver to detect objects about to enter the road.

It is important to minimise glare in order that the visual performance of drivers is not adversely affected. Therefore limitations are placed on the intensity of light emitted from luminaires at angles greater that 70° from the vertical. At lower angles, controls are not so important because the roofs of vehicles generally provide screening of the luminaires from the view of the driver.

4.11.4   Wet conditions

Wet conditions are a special case because the road-surface reflective properties change to become more reflective and less diffuse. This lengthens and intensifies the tail of the luminance patch, producing longitudinal bright patches towards the observer. Although the average luminance of the road surface increases in wet conditions there is a reduction in the driver's visual performance and ability to detect objects due to reduced luminance uniformity.

4.11.5 Considerations

The lighting installation should provide adequate levels of illumination to

- Allow drivers to detect the presence of small dark objects on the road surface.
- Allow drivers to detect pedestrians and or animals on areas immediately adjacent to the carriageway.

Lighting that achieves these primary aims is likely to be sufficient to fulfil the secondary aims of:

- Deterring criminal activity
- Enabling people and their intent to be recognised
- Enabling drivers to recognise pedestrians
- Enabling vehicles (including bicycles) to be recognised by pedestrians
- Enabling people not familiar with the area to recognise features such as a property name or number.

However, the lighting should be unobtrusive and well controlled. Whilst some light outside the boundary of the highway may be acceptable and beneficial, light entering into property, e.g. bedrooms is generally unacceptable. Maximum limits are shown in Table 2.2. The items of equipment, e.g. lighting columns and luminaires used in the installation should be in empathy with the general area and environment.

It is equally important to consider both daytime and night-time appearance of the installation.

4.11.6 Choice of lamp type

Section 3.2 gives the range of lamps available to the designer. The high pressure sodium, ceramic discharge metal halide and metal halide types are recommended for traffic routes through town and village centres and residential areas. The low pressure sodium type is no longer recommended as it does not introduce colour rendition to the right scene.

4.11.7 Choice of luminaire

The choice of luminaire for this type of application is vast with many variations as can be seen from Section 3.4. The chosen luminaire should

- Have good photometric control.
- Have low maintenance requirements, i.e. good I.P. rating.
- Be vandal resistant.
- Harmonise with the surroundings.

Luminaires are available with various fixing arrangements. In general, side-entry fixings are favoured for traffic areas although post mount is an option that is increasingly being used. The choice should ensure that the aesthetic appearance of the scheme is not sacrificed, e.g. the lighting column, bracket and luminaire should compliment one another and the location. Care should be taken when using post-mounted luminaires to ensure that the luminaire is of sufficient size to complement the height of the lighting column. Recommended bracket projections are shown in Table 4.25.

4.11.8 Supports

4.11.8.1 General

The method of fixing will normally be for the luminaire to be mounted on a lighting column. However, consideration should be given wherever possible to the use of adjacent buildings for wall-mounted luminaires.

4.11.8.2 Lighting columns

When choosing the lighting column care should be taken to ensure the following factors are considered:

- Scale and height should be in keeping with surrounding buildings, trees and other salient features in the field of view
- Siting should be in empathy with the surrounding features

Table 4.25 Recommended bracket projection for mounting height

Mounting height (m)	Recommended bracket projection (mm)	Absolute maximum bracket projection (mm)
8	500	1000
10	750	1500
12	1000	2000

- Lighting columns should be sited to minimise visual intrusion when seen from inside a dwelling
- Access to driveways and parking areas should be avoided
- Finished colour should be in empathy with the surrounding features
- Bracket projection should be in keeping with lighting column height (see Table 4.25)
- Brackets should compliment the design and proportions of the lighting column, luminaire and location
- Access for maintenance to ensure future repairs can be undertaken with minimum disruption (e.g. an opposite arrangement as opposed to twin central)
- Private electricity-distribution networks should be considered to remove reliance on electricity companies for repairs.

4.11.8.3   Proximity to carriageway

The closer lighting columns are to the carriageway the more likely they are to be damaged by vehicle impact and the more serious any damage is likely to be to vehicles and their occupants. Therefore it is recommended that lighting columns are not positioned at the toe of footways. The minimum and desirable clearance from the edge of the carriageway are shown in Table 4.26.

4.11.8.4   Wall mountings

This method of fixing should not be overlooked as it reduces the amount of street furniture in the field of view. A typical example is where a block of flats (two storey or more) is situated close to or directly to the rear of the road to be illuminated. Wall-mounted lighting can be particularly useful in city/town/village centres where there is a requirement to reduce visual as well as physical clutter and improve the daytime visual appeal of the area.

Table 4.26 Minimum and desirable clearances from the edge of the carriageway to the face of the lighting column

Design speed (km/h)	Design speed (mph)	Desirable clearance (m)	Minimum clearance (m)
50	30	1.5	0.8
80	40	1.5	1
100	50 and 60	2.0	1.5
120	70	2.0	1.5

The following should be taken into account.

- Permission from the building owner must be obtained.
- Wayleave agreements should be completed.
- Method of electrical connection.
- Access for maintenance.
- Aesthetic appearance during day and night-time.
- Avoidance of light pollution to adjacent dwellings.

4.11.9   Advantages of road lighting

The Government accept that the installation of road lighting on previously unlit roads reduces the night-time road accident rate by 31 per cent from that which it would otherwise be.[6] It allows travellers to orientate themselves, reduces pedestrian trip accidents and significantly reduces fear of crime. The government accepts that good street lighting reduces crime by a minimum of 20 per cent.[7] For details refer to Section 2.2.

4.11.10   Recommendations

Table 4.27 Recommended lighting requirements for ME traffic route classes

Area to be illuminated	EN 13201 class	Maintained average luminance (in cd m^{-2}) not less than	Uniformity ratio not less than	Longitudinal uniformity ratio not less than	Threshold increment (%) not greater than*	Minimum surround ratio**
Traffic routes	ME1	2.0	0.4	0.7	10	0.5
	ME2	1.5	0.4	0.7	10	0.5
	ME3a	1.0	0.4	0.7	15	0.5
	ME3b	1.0	0.4	0.6	15	0.5
	ME3c	1.0	0.4	0.5	15	0.5
	ME4a	0.75	0.4	0.6	15	0.5
	ME4b	0.75	0.4	0.5	15	0.5
	ME5	0.5	0.35	0.4	15	0.5
	ME6	0.3	0.35	0.4	15	Not required

Source: Reproduced by permission of BSI.

Notes

For S classes see Table 4.31 (in Section 4.12.6)

* Low pressure sodium lamps and fluorescent tubes are normally considered to be low luminance lamps. For these lamps and luminaires providing less or equivalent luminance, an increase of five percentage points in TI can be permitted.

** This criterion may be applied only where there are no traffic areas with their own requirements adjacent to the carriageway.

Table 4.28 Recommended lighting requirements for conflict class areas

Area to be illuminated	EN 13201 class	Maintained average illuminance (in lux) not less than	Maintained minimum illuminance (in lux) not less than	Uniformity ratio not less than	Associated traffic route lighting class
Conflict areas (including round-abouts)	CE0	50	20	0.4	ME1
	CE1	30	12	0.4	ME2
	CE2	20	8	0.4	ME3
	CE3	15	6	0.4	ME4
	CE4	10	4	0.4	ME5
	CE5	7.5	3	0.4	ME5

Source: Reproduced by permission of BSI.

Table 4.29 Recommended lighting classes for traffic routes

Category	Hierarchy description	Type of road general description	Detailed description
1	Motorway	Limited access Motorway regulations apply	Routes for fast-moving long-distance traffic. Fully grade separated and restrictions on use

Location	Feature (ADT)	Class
Main carriageway in complex interchange areas	≤40 000	ME1
	>40 000	ME1
Main carriageway with interchanges <3 km	≤40 000	ME2
	>40 000	ME1
Main carriageway with interchanges ≥3 km	≤40 000	ME2
	>40 000	ME2

Free flow link roads connecting motorways should be lit to the same standard as the main carriageway of the motorways they are connecting.

Motorway slip roads may be lit to one level lower than the main carriageway but retaining the same uniformity as the main carriageway. Slip road lighting should be extended to cover the full length of the slip road to provide additional lighting at the conflict point.

Hard shoulders	ME4a

Category	Hierarchy description	Type of road general description	Detailed description
2	Strategic route	Trunk and some principal 'A' roads between primary destinations	Routes for fast-moving long-distance traffic with little frontage access or pedestrian traffic. Speed limits are usually in excess of 40 mph and there are few junctions. Pedestrian crossings are either segregated or controlled and parked vehicles are usually prohibited

Location	Feature (ADT)	Class
Single carriageways	≤15 000	ME3a
	>15 000	ME2
Dual carriageways	≤15 000	ME3a
	>15 000	ME2

Conflict areas should be lit to the appropriate standard.

Table 4.29 (Continued)

Category	Hierarchy description	Type of road general description	Detailed description
3a	Main distributor	Major urban network and inter-primary links. Short–medium distance traffic	Routes between strategic routes and linking urban centres to the strategic network with limited frontage access. In urban areas speed limits are usually 40 mph or less, parking is restricted at peak times and there are positive measures for pedestrian safety reasons

Location	Feature (ADT)	Class
Single carriageways	≤ 15 000	ME3a
	> 15 000	ME2
Dual carriageways	≤ 15 000	ME3a
	> 15 000	ME2

In urban areas consideration should be given to the use of ME3b or ME3c in view of the lower speeds and shorter viewing distances.

Conflict areas should be lit to the appropriate standard.

Category	Hierarchy description	Type of road general description	Detailed description
3b	Secondary distributor	Classified road (B and C class) and unclassified urban bus routes carrying local traffic with frontage access and frequent junctions	In rural areas these roads link the larger villages and HGV generators to the strategic and main distributor network. In built-up areas these roads have 30 mph speed limits and very high levels of pedestrian activity with some crossing facilities including zebra crossings. On-street parking is generally unrestricted except for safety reasons

Location	Feature (ADT)	Class
Environmental zone E1/E2	≤ 7 000	ME4a
	> 7 000	ME3b
	≤ 15 000	
	> 15 000	ME3a
Environmental zone E3	≤ 7 000	ME3c
	> 7 000	ME3b
	≤ 15 000	
	> 15 000	ME2

Conflict areas should be lit to the appropriate standard.

Table 4.29 (Continued)

Category	Hierarchy description	Type of road general description	Detailed description
4	Link road	Road linking between the main and secondary distribution network with frontage access and frequent junctions	In rural areas these roads link the smaller villages to the distributor network. They are of varying width and not always capable of carrying two way traffic. In urban areas they are residential or industrial inter-connecting roads with 30-mph speed limits random pedestrian movements and uncontrolled parking

Location	Class
Environmental zone E1/E2	ME5
Environmental zone E3	ME4b or S2
Environmental zone E3 (urban) with high pedestrian or cyclist traffic	S1
Conflict areas should be lit to the appropriate standard.	

Source: Reproduced by permission of BSI.

Notes
1 Rural roads are classified as roads with a speed limit over 40 mph.
2 Urban roads are classified as roads with a speed limit up to and including 40 mph.

4.11.11 Vehicular tunnels

The major issue with tunnel lighting is the daytime situation, particularly in bright ambient conditions. When a vehicle driver approaches the tunnel portal, he will perceive a 'black hole', unless there is sufficient luminance inside the tunnel. This is due to the brightness contrast between the interior of the tunnel (its 'threshold' zone) and what surrounds it, such external surfaces as the roadway, the tunnel façade, the cutting walls and any visible sky (Figure 4.23) (the 'access' zone).

During the approach to and travel through the tunnel the driver's visual system will gradually adapt to the lower levels of illuminance and luminance inside the tunnel. Provided the tunnel is of sufficient length this gradual adaptation of the driver's visual system allows the levels of lighting in the tunnel to be gradually reduced by the provision of different lighting zones along the length of the tunnel.

The speed of the vehicle is a significant factor in determining zone lengths as the faster the vehicle the further into the tunnel the vehicle will have travelled before the driver's visual system has adapted. The human visual system takes longer to adapt from bright conditions to darker than from

Figure 4.23 Appearance of tunnel to approaching driver.

darker to brighter; therefore there is a need for higher illumination at the entrance of the tunnel than at the exit. However, care must be taken to ensure that on exiting the tunnel the driver whilst still in the 'parting' zone can still see vehicles just inside the tunnel (the 'exit' zone) to allow safe manoeuvring. The adaptation curve for vehicle speed of 40 mph is shown in Figure 4.24.

Both direct and reflected sunlight can inhibit vision at either the entrance or the exit. The immediately external area can be manipulated to reduce the external light level, by being in cut, or having tall road-side trees or specifically designed physical barriers to light. By day a very high level of lighting is needed in the threshold zone of the tunnel, typically up to 10 per cent of the value in the access zone.

The next section of a long tunnel is called the transition zone and further in there is the interior zone. These can be lit to lower levels, the luminance of the interior zone being comparable with that of a traffic route by night.

There is no need to provide special lighting in the exit zone, as sufficient stray daylight normally provides a luminance gradient acceptable for adaptation. However, a long two-way tunnel needs threshold and transition zones at each end, and these are often provided in both halves of a twin-tube tunnel, so that traffic can travel in both directions when one tube is closed for maintenance.

A programme of several levels of luminance in each of the zones is provided to allow for the natural changes in daylight, the lighting by night

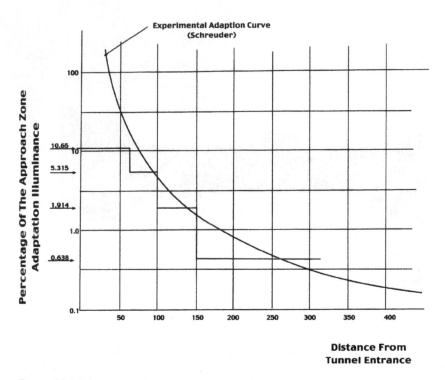

Figure 4.24 Adaptation curve for vehicle speed of 40 mph.

being constant throughout the tunnel and similar in quantity to that on the approach road. CIE Publication 88 (1990) 'Guide to the lighting of road tunnels and underpasses' gives recommendations on the distribution of luminances in time and space, depending on the speed limits within and approaching the tunnel.

In the threshold zone, multiple continuous or near-continuous rows of luminaires are commonly needed, but a single row usually suffices in the interior zone. It is important that spacings in this row are chosen so that vehicle drivers are not subjected to light pulses at frequencies within the epilepsy-triggering range of 2.5–15 per second, at the appropriate driving speed.

4.12   Residential areas

4.12.1   Introduction

The dictionary definition of residential states 'suitable for or occupied by dwellings', therefore people (some of whom may own vehicles) will occupy these areas. When planning lighting for such areas, at least as much emphasis should be given to people and their needs as to the requirements of vehicular traffic.

The prime needs of these people are that they feel safe and secure during the hours of darkness and that the environment where they live should be pleasant by day and by night.

4.12.2 Considerations

The lighting installation should provide adequate levels of illumination to

- Deter criminal activity
- Enable people and their intent to be recognised
- Enable drivers to detect pedestrians
- Enable vehicles (including bicycles) to be recognised by pedestrians
- Enable visitors to the area to recognise features such as a house name or number.

However, the lighting should be unobtrusive and well controlled. Whilst some light outside the boundary of the public area may be acceptable and beneficial, light entering into property, e.g. bedrooms is generally unacceptable. Maximum limits are shown in Table 2.2.

The items of equipment, e.g. columns and luminaires used in the installation should be in empathy with the general area and environment. It is equally important to consider both daytime and night-time appearance of the installation.

4.12.3 Choice of lamp type

Facial recognition is an important issue in residential area lighting. Therefore the choice of lamp is important, particularly with regard to colour rendition.

Chapter 3 gives the range of lamps available to the designer. The high pressure sodium, compact fluorescent and ceramic discharge metal halide types are recommended for residential areas. The low pressure sodium type is no longer recommended as it does not introduce colour rendition to the night scene.

4.12.4 Choice of luminaire

The choice of luminaire for this type of application is vast with many variations as can be seen from Section 3.4. The main consideration should be that the chosen luminaire should

- Harmonise with the surroundings
- Have low maintenance requirements, i.e. good I.P. rating

Table 4.30 Recommended bracket projection for mounting height

Mounting height (m)	Recommended bracket projection (mm)	Recommended maximum bracket projection (mm)
4	300	500
5	500	800
6	500	800

- Be vandal resistant
- Have good photometric control.

Luminaires are available with various fixing arrangements. In general post-mount, post-top and side-entry fixings are favoured for residential areas. The choice should ensure that the aesthetic appearance of the scheme is not sacrificed, e.g. the column, bracket, if used, and the luminaire should compliment one another and the location. Recommended bracket projections are shown in Table 4.30. However, a post-mount or post-top fixing of a luminaire with good optical control is recommended.

4.12.5 Supports

4.12.5.1 General

In residential areas the method of fixing would normally be for the luminaire to be column mounted. However, consideration should be given wherever possible to the use of adjacent buildings for wall-mounted luminaires.

4.12.5.2 Lighting columns

When choosing the lighting column care should be taken to ensure the following factors are considered:

- Scale and height should be in keeping with surrounding buildings, trees and other salient features in the field of view
- Siting should be in empathy with the surrounding features
- Lighting columns should be sited to minimise visual intrusion when seen from inside a dwelling
- Access to driveways and parking areas should be avoided
- Finished colour should be in empathy with the surrounding features
- Bracket projection should be in keeping with the lighting-column height (see Table 4.30)

- Brackets should compliment the design and proportions of the lighting column, luminaire and location (a lighting column with integral sweeping bracket may be preferable)
- Access for maintenance to ensure future repairs can be undertaken with minimum disruption (hinged lighting columns may need to be considered where vehicular access is limited or not available)
- Where a service strip exists the lighting column should be sited to the rear of it
- Where electricity supplies are to be taken direct from the electricity company's distribution system, the location of the electricity company cables should be ascertained to ensure wherever possible the most economic layout of lighting columns. Private electricity-distribution networks should be considered to remove reliance on electricity companies for repairs.

4.12.5.3   Wall mountings

This method of fixing should not be overlooked as it reduces the amount of street furniture in the field of view. A typical example is where a block of flats (two storey or more) is situated close to the area to be illuminated, possibly an adjacent parking area.

The following should be taken into account:

- Permission from the building owner must be obtained
- Wayleave agreements should be completed
- Method of electrical connection
- Access for maintenance
- Aesthetic appearance during daytime and night-time
- Avoidance of light pollution to adjacent dwellings.

4.12.6   Recommendations

BS EN 13201-2:2003 offers the following alternative means of designing and specifying residential lighting:

- S-series of lighting classes using horizontal illuminance
- A-series of lighting classes using hemispherical illuminance
- ES-series of lighting classes using semi-cylindrical illuminance
- EV-series of lighting classes using vertical plane illuminance.

Each of these classes of illumination is based on the quantity and quality of illumination to be provided.

BS 5489-1:2003 recommends the use of the S-series of lighting in the United Kingdom as these are more easily computed and measured using the same processes as previously used.

The recommended S-series lighting levels are shown in Table 4.31. BS EN 13201 makes no reference to and gives no advice as to how the correct lighting class for a residential road should be chosen; however, full information is provided in BS 5489-1:2003 allowing the correct class of illumination to be determined from the amount of night-time use by pedestrian and vehicular traffic, the crime rate and the environmental zone in which the area is situated.

Table 4.32 sets out a matrix using traffic flows, crime levels and the environmental zone to assist in the determination of the correct lighting class.

4.12.7 Footpaths

Many residential areas are built to minimise the intrusion of vehicular traffic and, therefore, to facilitate pedestrian movement and access, footpaths are a main feature. Consideration for lighting these footpaths, whilst similar to that for areas of mixed traffic (vehicular and pedestrian), do need taking other features into account.

The width of the footpath is likely to be much less than the areas of mixed traffic; therefore the use of luminaires with optical control specifically designed for footpaths may be considered. Whilst this may be the better option for efficiency, illuminating a wider area may be preferred to provide anyone using the footpath with a wider area of view. This may help reduce the fear of crime to the public. However, care should be taken to ensure that light does not intrude into adjacent properties.

Footpaths are generally prone to vandalism due to the remoteness and possible readily available 'ammunition' in the form of stones etc. The use of vandal-resistant materials is, therefore, of prime need, particularly the

Table 4.31 Recommended lighting classes for subsidiary roads and pedestrian areas

Area to be illuminated	EN 13201 class	Maintained average illuminance (in lux) not less than	Maintained minimum point illuminance (in lux) not less than	Maintained maximum average illuminance (in lux) not more than
Subsidiary roads and pedestrian areas	S1	15.0	5.0	22.5
	S2	10.0	3.0	15.0
	S3	7.5	1.5	11.25
	S4	5.0	1.0	7.5
	S5	3.0	0.6	4.5
	S6	2.0	0.6	3.0

Source: Reproduced by permission of BSI.

Table 4.32 Lighting recommendations for British subsidiary roads and pedestrian areas

Traffic flow including pedestrians and cyclists	Low		Normal		High	
Environmental zone	E1/E2	E3/E4	E1/E2	E3/E4	E1/E2	E3/E4
Low crime ($R_a < 60$)	S5	S4	S4	S3	S3	S2
Normal crime ($R_a < 60$)	S4	S3	S3	S2	N/A	S1
High crime ($R_a < 60$)	S2	S2	S2	S1	N/A	S1

Source: Reproduced by permission of BSI.

Notes
1 For the purpose of this table all roads are treated as allowing for parking.
2 Where a light source with $R_a >= 60$ is used, the lighting level may be reduced by one lighting class.
3 Low traffic flow refers to areas where the traffic usage is of a level equivalent to a residential road and solely associated with the adjacent properties.
4 Normal traffic flow refers to areas where the traffic usage is of a level equivalent to a housing estate access road and may be associated with local amenities such as clubs, shopping facilities, public houses, etc.
5 High traffic flow refers to areas where the traffic usage is high and may be associated with local amenities such as clubs, shopping facilities, public houses, etc.
6 Crime rates are relative to the local area, not national. Assistance should be obtained from the local crime-prevention officer.

luminaire bowl and column base compartment door. In areas known to be prone to vandalism the use of compartments near the top of the column should be considered.

Access for both installation and subsequent maintenance is also a prime consideration and the use of hinged typed columns should be investigated at the time of design. Many footpaths are constructed adjacent to landscaped areas and if possible, input at the design stage to ensure future tree/shrub growth does not interfere with the luminaire output, is desirable. If this is not possible arrangements to ensure routine pruning should be put in place.

4.12.8 Cycle tracks

Cycle tracks or cycle routes are part of the Government's strategy for assisting with environmental improvements by reducing the number of short journeys by car. This requires that cycle tracks are used around the clock, and hence the need for adequate lighting during the hours of darkness.

The following points are to be considered:

* The hierarchy of cycle tracks.
* Does the track require to be illuminated?
* What are the surrounding landscape features?
* Does the track have mixed pedestrian and cycle traffic?

- Maintenance of equipment once installed.
- Availability of electricity supply.
- Avoidance of spill light onto adjacent private property.

Further information is given in the Institution Technical Report No. 23, Lighting of cycle tracks.

4.12.9 Traffic calming

4.12.9.1 General

Various methods of reducing traffic speeds in residential areas are being used, and on a more frequent basis. By their nature these require the placing of objects in or on the carriageway to impede the progress of vehicular traffic.

Recommendations are based on lighting-performance measures specified in BS EN 13201. Further information is given in the Institution Technical Report No. 25, Lighting for traffic calming features.

4.12.9.2 Colour

The choice of light source for traffic calming features should provide the degree of colour rendition necessary for driver navigation and pedestrian orientation. Generally this will be provided by high pressure sodium lamps.

The low pressure sodium lamp is not recommended for the illumination of traffic calming features as it does not introduce colour rendition to the night scene.

If coloured surfacing is used to improve conspicuity of a traffic calming feature, it is important that the lamp used will reveal the colour at night, and the use of compact fluorescent or ceramic discharge metal halide lamps should be considered for this purpose.

4.12.9.3 Horizontal deflections

Generally, as one direction of traffic is given priority over the other, these features may be considered as conflict areas as described within BS EN 13201 and lit accordingly.

The choice of lighting level for the feature should be as follows:

- For roads lit using the CE series of lighting classes, one class higher than that of the road with a minimum level of lighting of CE4.
- For roads lit using the ME classes of lighting with normal viewing distances (greater than 60 m), one class higher than that of the through road. Care should be taken when selecting the appropriate ME lighting

class to ensure that the uniformity levels of the through road and the traffic calming feature are maintained.

- For roads lit using the ME classes of lighting with short viewing distances (less than 60 m), the CE class of lighting one class higher than that of the through road, with a minimum level of lighting of CE4.
- For roads lit using the S classes of lighting, one class higher than that of the road with a minimum level of lighting of S4.

The recommended lighting class for horizonal traffic calming features are tabulated in Table 4.33.

A change in light source to that used for the road is recommended as a method of drawing attention to the feature.

4.12.9.4 Vertical deflections

For round humps and cushions no specific arrangement is recommended; however, the lighting level should be increased by one class.

Where flat road humps are used to form a crossing place for pedestrians, the lighting should be such as to ensure drivers can see pedestrians on and in the vicinity of the crossing point. The area of the road hump and the adjacent footway(s) should be treated as a conflict area and lit using either the CE or S-series of lighting depending on the category of road. The lighting level to the feature should be increased by one lighting class with a minimum level of CE4 for roads lit using the CE series of lighting classes and S4 for roads lit using the S series of lighting classes. For flat humps, columns should not be located over the feature as this decreases conspicuity.

4.12.10 Parking areas

For economic reasons it is often the case that the maximum number of dwellings allowed by planning constraints are located within the available

Table 4.33 Recommended lighting class for horizontal traffic calming features

Lighting class series		Requirements	
Through road	Horizontal traffic calming feature	Additional	Minimum
CE	CE	One class higher	CE4
ME, Short viewing distance (<60 m)	CE	One class higher	CE4
ME, Normal viewing distance (>60 m)	ME	One class higher	
S	S	One class higher	S4

land space. This can mean that townhouses or flats are built without dedicated parking for vehicles owned by the occupants. In such cases defined parking areas are provided off the carriageway and often not adjacent to the properties.

It is important to provide adequate lighting for these areas, certainly to illumination levels equal to the adjacent carriageway.

The following points should be considered when lighting parking areas:

- Colour rendering of the light source
- Can the area be adequately illuminated using conventional road lighting techniques or should area lighting methods be employed?
- Are there adjacent buildings (such as multi-storey flats) to enable floodlights to be mounted at high level?
- Is access for future maintenance possible?

4.12.11 Pedestrian underpasses

Pedestrian underpasses should be pleasant in appearance, safe and clean and give an impression of security at all times. During daylight they should have sufficient illumination just inside the entrance to overcome the 'black hole' effect allowing the pedestrians' eyes to adjust to the change of illumination from a bright sunlit day to the much lower levels within the underpass. An increase in the level of illumination at the entrance of an underpass will help pedestrians see into the underpass, reducing their anxiety and assuring them that it is safe to enter. Short underpasses with 'see through', i.e. where the exit can be clearly seen from the entrance and makes up a substantial proportion of the visual scene, may not need increased lighting at the entrances.

Night-time raises alternative concerns for the underpass user. As well as requiring the underpass to be safe and secure the night-time user will require the approach ramps and stairs to be adequately lit. The entrance of the underpass should be fully visible from any external stairs and ramps allowing the user to see into the underpass before entering. Similarly a user exiting an underpass will wish to ensure that they can see outside the underpass before leaving it.

Stairs and ramps that are at right angles to the direction of the underpass are not recommended unless they are located remote from the underpass and can be seen before exiting to allow the pedestrian to orientate themselves and assess any potential dangers. Even so, such stairs and ramps should be clearly visible with open railings instead of solid obstructing walls which block the view of anyone hiding behind. A long, gentle sloping footpath approaching the underpass in the same general direction as the underpass is the preferred option for ingress and egress. The exit to the underpass should have sufficient illumination at night to allow the pedestrian to see outside without the 'black hole effect', and the illumination should be gradually tapered down to that of the normal street lighting of the area. Abrupt transitions from the high levels of the interior of the underpass to the lower external levels of illumination are not recommended. Areas of vegetation on the approach slopes of underpasses should be kept well trimmed to reduce the chance of anyone hiding in them and should have sufficient illumination at night to reveal anyone in them.

It is recommended that the interior lighting of the underpass be provided by lamps with a full spectrum to enhance the visual scene and provide a sense of comfort and familiarity.

Luminaires and associated electrical equipment should be vandal resistant and sealed to a minimum of IP5x; however, where power water jet cleaning is envisaged the sealing should be increased to a minimum of IP6x.

Luminaries should be mounted as high in the walls as possible or be cornice mounted. The mounting of luminaires flush into the ceiling is not recommended as they do not illuminate the ceiling giving the impression of a reduction in height. Semi-recessed ceiling luminaires give more light onto the ceiling but offer a greater area to vandal attack.

Cornice lighting mounted at the junction of the wall and ceiling offer good lighting on the walls, ceiling and users. Such luminaires can be installed in continuous lengths along the length of the underpass and can be used to accommodate the electrical wiring. If continuous lengths of cornice luminaires are not used it is recommended that infill panels be used between the luminaires to present a neater continuous row and also to protect the ends of the luminaires from vandal damage. This arrangement will also allow the electrical wiring to be safely housed. If the underpass is wider than 6 m then it is recommended that two rows of cornice units be used; alternatively, two rows of ceiling mounted luminaires may also be considered.

The lighting in long underpasses should be wired in separate circuits to allow for switching and duplication in case of damage or failure. Allowance should be made to reduce lighting levels during the night to save energy and reduce visual impact. Similarly consideration can be given to the use of luminance meters to switch additional lighting at the entrance of tunnels during periods of high brightness; however, unless the underpass is extremely complex and well used the additional cost over that of normal day/night switching probably cannot be justified. The switching of lighting in short underpasses can generally be restricted to simple day/night settings.

Underpasses should be regularly inspected and maintained with graffiti and rocks and stones being removed to reduce the impact of vandalism. Well-maintained and inviting underpasses will be used more regularly than dirty, dingy smelly ones. Recommendations for the lighting of subways, footbridges, stairways and ramps are given in Table 4.34 and recommendations for the lighting of subways, footbridges, stairways and ramps in residential areas are given in Table 4.35.

Table 4.34 Lighting recommendations for subways, footbridges, stairways and ramps

Type	Day		Night	
	Maintained average illuminance (in lux) not less than	*Maintained minimum point illuminance (in lux) not less than*	*Maintained average illuminance (in lux) not less than*	*Maintained minimum point illuminance (in lux) not less than*
Subways				
Open	N/A	N/A	50	25
Enclosed	350	150	100	50
Footbridges				
Open	N/A	N/A	30	15
Enclosed	350	150	100	50
Stairway/ramp				
Open	N/A	N/A	30	15
Enclosed	350	150	100	50

Source: Reproduced by permission of BSI.

Table 4.35 Lighting recommendations for subways, footbridges, stairways and ramps located in residential areas

Type	Day		Night	
	Maintained average illuminance (in lux) not less than	*Maintained minimum point illuminance (in lux) not less than*	*Maintained average illuminance (in lux) not less than*	*Maintained minimum point illuminance (in lux) not less than*
Subways				
Open	N/A	N/A	30	15
Enclosed	100	50	50	25
Footbridges				
Open	N/A	N/A	30	15
Enclosed	100	50	50	25
Stairway/ramp				
Open	N/A	N/A	30	15
Enclosed	100	50	50	25

Notes on Tables 4.34 and 4.35
1 'Open' equates to major daylight penetration and will include the majority of subways in residential areas.
2 For 'enclosed' areas emergency lighting needs to be considered. It is essential that it is installed if the area forms part of an escape route from a shopping centre, car park or transport interchange.
3 Where longer subways have poor daylight penetration, or where subway user confidence needs to be ensured, it may be necessary for the threshold illuminance value to be up to twice the value of the general daytime service level.
4 The recommendations for stairways and ramps do not include those giving exterior access for disabled people to buildings. Reference should be made to BS 8300.

4.13 Car parks[8]

4.13.1 General

Car parks vary from small, private car parks in rural areas serving one small company through to large commercially operated public car parks in the centre of a town or city. Many of these car parks, particularly those situated in town centres and extensively used by the public during the hours of darkness, would benefit from the provision of lighting which would improve the safety and security of the clientele and their property.

The improvements that can be gained in safety and security by the provision of suitable lighting are recognised and promoted by the Association of Chief Police Officers (ACPO) in their Secured Car Park Award Scheme which requires car parks to reach predefined standards of physical design and management. The provision, maintenance and operation of a system of lighting in accordance with recognised standards are a major prerequisite of an award under this scheme.

4.13.2 Environmental issues

Care should be taken when designing, installing and operating car park lighting to ensure that excessive light does not spill beyond the boundary of the car park on to adjacent private property. Levels of spill light, glare and sky glow should not exceed those recommended in Section 2.3 for the specific environmental zone in which the car park is situated.

4.13.3 Choice of equipment

The type of equipment used to light car parks should be carefully chosen to provide the correct levels of illumination whilst taking account of the character of the surrounding area and buildings. Low mounting heights of 5 and 6 m can help to reduce the visual intrusion of the installation against the skyline, particularly by day. However, lighting systems designed using low mounting heights can result in a proliferation of lighting columns, increasing capital and operational costs and poor and wasteful installations unless careful consideration is given to the choice of luminaires used and the design and installation of the system.

Lighting systems using higher mounting heights of up to 12 m can provide a balance between excessive numbers of lighting columns and visual intrusion whilst allowing an economic and environmentally acceptable lighting solution to be provided.

Many car parks are lit using floodlights mounted on the top of lighting columns. Such installations can result in excessive glare, poor uniformity and excessive overspill of light. To minimise glare and overspill the main-beam angle of the floodlights should be kept below 70° from the vertical.

The use of higher mounting heights will allow lower main-beam angles to be used, reducing glare and overspill whilst still illuminating the same area. In environmentally sensitive areas such as environmental zones E1 and E2 and where light intrusion and overspill may cause a problem, the use of double asymmetric beam floodlights designed to allow their front glazing to be kept at or near parallel to the car park surface is recommended.

Alternatively, road lighting luminaires fitted with a flat glass and mounted horizontally are also suitable for use in such areas. However, it may still be necessary in particularly sensitive locations to further screen floodlights and luminaires to reduce overspill and intrusive light into adjacent properties.

On large car parks the use of luminaires, or clusters of luminaires, which provide a horizontal 360° symmetrical light output may be beneficial in reducing the number of lighting units installed making for a cost-effective lighting system. Many of these luminaires are designed to be fitted with internal screens to reduce light emitted in a particular direction without distracting from the general appearance of the luminaire. This feature makes this type of luminaire effective in controlling overspill light and glare whilst maintaining the appearance of the installation.

High masts are generally not used to light car parks due to the daytime appearance of the installation. However, such equipment can be used with advantage to reduce the number and clutter caused by lower mounting height columns. The use of high masts can also reduce the number of parking bays lost or reduced in size due to the positioning of the lighting columns. Where high masts are used for this purpose, care must be taken when designing the scheme to allow access for maintenance, particularly if the columns are base hinged and lowered to the ground for maintenance.

4.13.4   Light sources

General car park lighting should be provided by lamps with an $R_a > 20$. However, in areas of high amenity, consideration should be given to the use of lamps with an $R_a > 65$. The use of low pressure sodium lamps for lighting car parks is not recommended due to the lack of colour rendering, making the recognition of cars difficult.

Where CCTV is to be used to control and manage traffic, lamps with an $R_a > 20$ will be suitable; however, if it is intended to use the CCTV for the detection of crime and the recognition of criminals for prosecution it is recommended that lamps with an $R_a > 80$ be used.

4.13.5   Design considerations

4.13.5.1   General

Where lighting of a car park is to be provided by floodlights mounted on lighting columns, these are best located around the perimeter of the car park where they will cause the least obstruction to car parking spaces. With careful selection of the floodlight to be used and careful design of the installation, the effects of the lighting outside the area to be lit can be minimised. In large car parks it is not always possible to provide lighting only from the perimeter of the area. In such situations it may be necessary to provide additional lighting columns within the car parking area. Where this is necessary the use of multiple road lighting luminaires mounted on lighting columns or special luminaires which provide a horizontal 360° symmetrical distribution can provide a satisfactory and economical solution.

Care and consideration should be given to access to the equipment for maintenance and repair. Lighting columns should, wherever possible, be erected in traffic islands or garden areas where they are protected from collision damage. This is not always possible on car parks where such features are limited to maximise the number of car parking spaces provided. In such situations it is recommended that the lighting columns be installed at the intersection of four parking bays where they will cause least disruption and reduction of space. The provision of additional protection around the base of the lighting column to protect them from collision damage is recommended. This is particularly necessary on parking areas used by heavy goods vehicles (HGV) and commercial vehicles where the driver has limited vision when reversing.

4.13.5.2 HGV and high-sided vehicle parking areas

In general the recommendations made in this section refer to both car parks and parking areas specifically used by HGV and high-sided commercial vehicles. However, the following details are specifically related to the latter. The use of low mounting height lighting columns for lighting systems on vehicle parking areas extensively used by HGV and high-sided commercial vehicles is not recommended as light is unable to penetrate between adjacent parked vehicles. In such situations the use of 12-m mounting height columns as a minimum is recommended with care being taken in the design of the system to ensure that the sides of vehicles are adequately lit with no undue shadows. This may entail a reduction of the optimum design spacing to ensure adequate overlap and penetration of light between adjacent tall vehicles, particularly if directional floodlighting is being used. It is recommended that a detailed study of the use or the proposed use of the vehicle park is made to establish the general location, direction and spacing of HGV and high-sided vehicles to determine the possible areas of shadow.

4.13.6 Hours of operation

Many public car parks are open and operated 24 hours a day thus requiring the provision of lighting throughout the night. However, a general reduction of lighting over the full area of the car park by switching off some of the lights, or by dimming all of the lighting system, is recommended whenever car use reduces significantly below the park capacity.

Where a car park is locked with no public access for part of the night then it is recommended at least some of the lights are turned off during this period, although consideration should be given to the protection of any cars left on the park. If a car park is heavily used during the day with only a low level of use during the hours of darkness it may be possible to reserve one area for 24 hour operation, whilst lighting in the remainder is extinguished. The restriction on levels of lighting and hours of operation not only reduces the operational costs of the system but helps to reduce the possible detrimental effect of the lighting on the environment and the adjacent area and properties.

Many small private car parks, particularly those associated with businesses which operate fixed opening hours, are not used at all for a large period of the night. In such situations it is recommended the lighting is turned off after the last workers' cars have left the park, unless it is required for building security.

4.13.7 Maintenance

To ensure that the lighting on a car park remains effective and fulfils the design brief, an efficient maintenance system must be implemented. Further details of how maintenance procedures, systems and frequencies are developed are given in Section 3.7.

4.13.8 Recommendations

The standards of lighting for car parks recommended in Table 4.36 recognise the needs of the users whilst taking account of the detrimental effects exterior lighting may have on the environment if not correctly specified, designed, installed and maintained. Allowances for varying environmental factors are made by recommending different levels of illumination according to the location, size and use of the car park.

4.14 Security lighting

4.14.1 General

The rates of burglary and theft from business premises and factories clearly demonstrate the need for good security of property and premises. Security lighting on its own can, because of its deterrent effect, be very successful. However, it can also be used very successfully as an integral part of an overall package including walls, fences and security checks.

Table 4.36 Recommended lighting levels for car parks

Location and lighting class	Maintained average illuminance (in lux) not less than	Overall uniformity not less than
Rural (environmental zones E1 and E2)		
Large car park		
Short stay	15	0.4
Long stay	15	0.4
Small car park		
Short stay	10	0.4
Long stay	10	0.4
Local car park	5	0.4
Urban (environmental zones E3 and E4)		
Large car park		
Short stay	30	0.4
Long stay	15	0.4
Small car park		
Short stay	20	0.4
Long stay	15	0.4
Local car park	10	0.4

Notes
1 For zone descriptions see Section 2.3.2.
2 Where the car park lies on the boundary of two zones or can be observed from another zone, the obtrusive light limitation values used should be those applicable to the most rigorous zone. It is recommended that consultations take place with the local planning authority to determine the exact environmental zone to be used for the specific location.
3 For car park definitions see Section 4.13.6.

4.14.2 Environmental issues

All security lighting should be suitably screened to avoid annoyance to road users or adjacent property holders and to protect the environment from both light trespass and light pollution. Additional information is given in Section 2.3: Light Pollution.

4.14.3 Choice of equipment

The type of equipment used to provide security lighting should be carefully chosen to provide the correct levels of illumination whilst taking account of the character of the surrounding area and buildings. Lower mounting heights of 5 and 6 m can help to reduce the visual intrusion of the installation against the skyline, particularly by day. However, lighting systems designed using lower mounting heights can result in a proliferation of lighting columns, increasing capital and operational costs and resulting in poor and wasteful installations, unless careful consideration is given to the choice of luminaires used and the design and installation of the system.

In sensitive and built-up areas it may be necessary to fit baffles to further restrict light trespass and light pollution. Floodlights and other adjustable luminaires should be fitted with a positive locking system to allow the luminaire to be locked in position after being adjusted and aimed.

In environmentally sensitive areas such as environmental zones E1 and E2 (see Section 2.3.2) and where light intrusion and overspill may cause a problem, the use of double asymmetric beam floodlights (see Figure A.9C) designed to allow their front glazing to be kept at or near horizontal is recommended. Alternatively, road lighting luminaires fitted with a flat glass and mounted horizontally are suitable for use in such areas. However, it may still be necessary in particularly sensitive locations to further screen floodlights and luminaires to reduce overspill and intrusive light into adjacent properties.

Where luminaires are mounted on lighting columns care must be taken to ensure that they are accessible for maintenance using a mobile access maintenance platform. Alternatively, the columns should be base or mid hinged, or, fitted with raising and lowering gear to bring the luminaires down to ground level for maintenance.

4.14.4   Light sources

The ideal lamp for security lighting is one which gives full light output as soon as switched, has a long life, is economical to operate and comes in a

Table 4.37 Light sources for security lighting

Lamp type	Colour rendering index (R_a)	Colour appearance	Security lighting use
Low pressure sodium	0	Orange	Not recommended
High pressure sodium	20	Golden yellow	General security lighting
High pressure mercury	40–50	Bluish white	General security lighting
Metal halide	65–90	Crisp white	General security lighting
Fluorescent and compact fluorescent	80–90	White – in varying hues	General and domestic security lighting
Tungsten halogen	100	Crisp white	Domestic security lighting

Note
For light sources for CCTV see Section 4.3.3.3. Full details of light sources are given in Section 3.3.

range of sizes to cover all types of application. Unfortunately the ideal lamp does not exist. However, high pressure sodium, metal halide and tubular and compact fluorescent lamps provide a practical and economical choice (Table 4.37).

4.14.5 Design considerations

The basic principles of security lighting are

- To provide illumination to assist the detection of intruders
- To avoid shadows which might offer concealment
- To deter an intruder by creating an environment of potential exposure.

A virtue of security lighting is that it seldom needs to perform only one function. It has the ability to provide lighting for amenity, pleasure, display, advertising or even work and storage. This multiplicity of uses can increase its cost-effectiveness, and with careful design the security aspects may be provided at little or no additional cost. If a building stands in its own grounds then the lighting system can be designed to provide security, car parking and general amenity lighting as well as advertising the presence of the occupier.

The type of property, its use and the area in which it stands must be considered. A small property containing low-value items situated in a well-lit, well-observed area such as a city centre will have little attraction to a criminal. However, moving that same property to a poorly lit area on the outskirts of the town or city may increase the interest of the criminal. Similarly, if the value of the product manufactured or stored at the property is increased so will be the interest of the criminal. When determining the potential for criminal activity the entire location of the property must be taken in to account. From the front the property may be situated on a busy well-lit city centre street, whereas the rear of the property may be poorly or not lit at all with very little natural observation. In this situation it may not be necessary to provide any security lighting to the front of the property; however, the rear of the property will probably require a system of security lighting.

An important feature of security lighting is to make things appear to be bright. This does not necessarily mean that large quantities of light have to be provided. It is often possible to simply direct light towards the wall of a building so that an intruder will be seen either as a lit figure or as a silhouette against the bright building depending on which side of the lighting fittings he is standing. Alternatively if a high fence or wall surrounds the property then the exterior of this may similarly be lit to provide a bright background against which intruders may be observed. This system of security lighting is particularly applicable where passer-by or visiting security patrols are relied upon to observe and report possible intruders. If the fence or wall can be made from light coloured materials or painted in a light colour then the effectiveness of this system will be greatly improved.

4.14.6 Lighting to deter

It has been common practice to provide security lighting which is attached to the building to be protected, or which consists of free-standing lighting units directed to shine light outwards. These systems are designed to cause discomfort, confusion and uncertainty to potential intruders whilst providing any security guards patrolling and concealed behind the security lighting with a good view of the possible intruder. Such systems are still used in many military security lighting situations. However, they are losing favour for general security lighting due to the adverse effect they have on the visual amenity and the environment of the surrounding properties and the area.

4.14.7 Lighting to reveal

Many businesses have an enclosed yard or loading bay to the side or rear of the property. With careful design it should be possible to position luminaires to serve not only as security lighting but also as functional lighting for loading and moving goods. As the general level of illumination required for security may be less than that required for functional lighting, differential switching may be used to control the levels of lighting provided. This could be achieved by a time switch or photocell switching the security lighting with additional functional lighting being manually switched as and when required. Thus lighting is provided for security and this can be boosted for working under as and when needed, saving energy and reducing light pollution. Again, it is advantageous if the yard or loading bay is made as light in colour as possible; a concrete surface and white walls having a high reflectance not only makes maximum use of the available light, but provides better contrast for a dark figure or its shadow.

Where a doorway opens on to a well-lit street it may not be necessary to provide additional lighting as the street lighting may provide adequate coverage. However, care should be taken to ensure that the street lighting is operational all night, and consideration should be given as to what would happen if it was out of operation due to a fault.

Where there is no street lighting or the door to the property is recessed then a security light should be mounted above the door in a position that is screened from the inside of the property but provides adequate illumination to the face and body of a person standing at the door. Additional lighting may also be provided above the door opening on the outside of the property, but this should only be done when it will not cause a shadow in the door recess or when additional lighting has been provided above the door inside the recess. Care should be taken to ensure that any lighting units are beyond the reach of a criminal. Preferably they should be incorporated into a general or display lighting system.

4.14.8 Open spaces

Open spaces such as car parks, recreation grounds and school playgrounds may well attract criminals and vandals but lighting, by its deterrent effect, can give a measure of protection. Many of these areas are now provided with a system of lighting to extend the use of the area to after dark. By differential switching, a part of such a system could be left in lighting throughout the night to provide security.

4.14.9 Hours of operation

The purpose of security lighting is to try and improve the security of the building or property being protected. It is therefore essential that security lighting be provided throughout the entire hours of darkness.

4.14.10 Security of supply

The chance of an opportunist taking advantage of a power supply failure is a very small risk and would generally be accepted unless the property being protected has a very high value or extreme security is required. In such situations the provision of a standby system such as batteries and inverters or by a self-contained stand-by generator set should be considered. It is possible to provide a no-break system that will provide an instantaneous changeover without extinguishing any discharge lamps that may be in use.

4.14.11 Installation methods

Good security lighting is a requirement of many companies and property owners. Its value and effectiveness can, however, be reduced if care and consideration is not given to its installation and location.

Luminaires fitted to the exterior walls of properties should be installed out of reach of any criminal or vandal. Where the wall is easily accessed by vehicles further consideration should be given to ensure that the luminaire is out of reach of someone standing on the top of a large vehicle.

The electrical wiring to the security lighting should preferably be run where it cannot be interfered with. Power supplies can be run inside the building for protection from both the elements and miscreants. Where this cannot be achieved the best protection is to place them out of reach or to house them in steel conduit.

For free-standing exterior lighting the power supply is best provided by armoured cable in plastic ducting buried a minimum of 500 mm in the ground within the site.

4.14.12 Maintenance

The routine cleaning and lamp replacement of lighting installations is dealt with in Section 3.7. However, due to its nature, security lighting has some special maintenance requirements.

Any broken lamps or luminaires should be regarded with suspicion. The breakages may be accidental or they may be the start of a systematic attempt to put the system partially out of operation. Any signs of damage or lamp failures should be reported and dealt with immediately. Any signs of attempted interference with the installation or the luminaires should be investigated and, in the case of an exterior floodlighting installation, any sudden misalignment of the luminaires should receive similar consideration and prompt correction.

Dependent upon the type of equipment used and the availability of suitably trained staff it may be practical to maintain a small stock of spares such as the front glass of the luminaire, lamps and control gear to allow easy and quick repairs to be carried out.

4.14.13 Recommendations

The recommendations for security lighting recognise the needs of the users whilst taking account of the detrimental effects exterior lighting may have on the environment if not correctly specified, designed, installed and maintained (Table 4.38). Allowances for varying environmental factors are made by recommending different levels of illumination according to the location, size and use of the area.

4.14.14 Domestic security lighting

Well designed, installed and maintained domestic security lights bring comfort and well being to our lives providing us with a sense of security in our homes. Because of their price and ease of installation, tungsten–halogen floodlights are often used for this purpose. These units can provide satisfactory security lighting if correctly installed and aimed. However, it is rarely necessary to use a lamp of greater than 150 W. Movement detectors to sense the movement of intruders can be useful if they are correctly installed and aimed.

Table 4.38 Recommended lighting levels for general area security lighting

Location	Large area		Small area and domestic security lighting	
	Maintained average illuminance (in lux) *not less than*	Overall uniformity *not less than*	Maintained average illuminance (in lux) *not less than*	Overall uniformity *not less than*
Rural (Environmental zones E1 and E2)	10	0.4	5	0.4
Urban (Environmental zones E3 and E4)	20	0.4	15	0.4

Notes
1 For zone descriptions see Section 2.3.2.
2 Where the car park lies on the boundary of two zones or can be observed from another zone, the obtrusive light limitation values used should be those applicable to the most rigorous zone. It is recommended that consultations take place with the local planning authority to determine the exact environmental zone to be used for the specific location.
3 Large areas are considered to be in excess of 500 m².
4 For perimeter lighting a minimum vertical illuminance of 1 lux maintained should be provided at the distance and height at which detection is required.

Floodlights and detectors should be aimed to only detect and light people on your property. They should not detect a person or animal walking down the street. If the floodlight is fitted with a timer, this should be adjusted to the minimum to reduce the operation of the light.

For many properties, a better solution for domestic security lighting is to use a bulkhead or porch light fitted with a 9- or 11-W compact fluorescent lamp. These units can be left lit all night, providing all night security, for only a few pounds of electricity per year. For further details on domestic security lighting see the ILE leaflet 'Domestic Security Lighting – Friend or Foe?' which can be downloaded from the ILE website at www.ile.org.uk.

Appendix

A.1   General

A.1.1   Inverse square law

The 'Inverse square law' states that the illuminance will be the intensity of a light beam divided by the aiming distance squared (Figure A.1). In simple terms, doubling the distance from light source to target means achieving one-quarter of the illuminance on a surface.

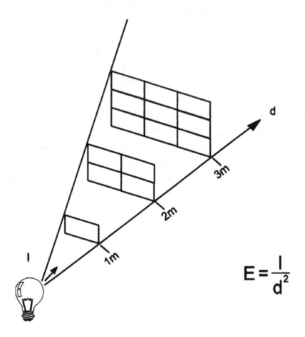

$$E = \frac{I}{d^2}$$

Figure A.1 Inverse square law.

A.1.2 Cosine law

The 'Cosine law' states that the illuminance is proportional to the cosine of the angle of incidence of the intensity of a light beam (Figure A.2). The angle of incidence is measured from the normal (the perpendicular) to the surface at the incident point. Light beams arriving vertically produce the largest illuminance because the cosine of 0° is 1, that is the maximum possible value. The effect is due to the intensity landing on a larger surface area when it is incident at an angle.

A.1.3 Environmental zones

Environmental zones are a recognised way of separating general areas to which different lighting criteria should apply. Environmental zone classifications are shown in Table 2.1.

For more information on the use of environmental zones related to light pollution, see Section 2.3.

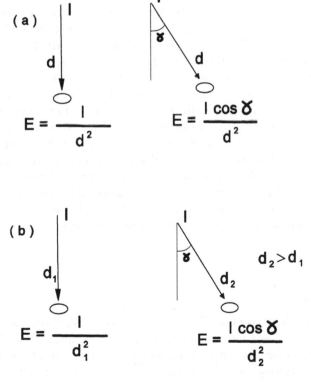

Figure A.2 Cosine law.

A.1.4 Lighting design

A.1.4.1 General

Lighting design is a complex process, with no hard and fast rules that will suit all design problems for every designer. Nevertheless, the following design approach represents good practice and will give guidance to less-experienced designers.

A.1.4.2 Objectives

The first stage in planning any lighting design is to establish the lighting design objectives. These objectives can be considered in three parts as follows.

First, the lighting must be safe in itself and allow personnel to work and move around in safety. It is therefore necessary to identify any hazards present and to consider the use of emergency lighting.

Second, the type of work which takes place in the area will define the nature and variety of visual tasks. An analysis of these visual tasks in terms of size, contrast, duration, need for colour discrimination and so on is essential to establish the quantity and quality of lighting required to achieve satisfactory visual conditions. In addition to establishing the nature of tasks to be done, it is also necessary to identify the position where the tasks occur, the planes in which they lie and the extent of any obstructions. This information is essential if lighting matched to the tasks is to be provided.

Finally, the lighting of a space will affect its character, and the character of objects within it. It is therefore necessary to establish what mood or atmosphere is to be created. When establishing the objectives, it is important to differentiate between those that are essential and those that are desirable. It is also important to establish both the design objectives and the design constraints. There are many constraints that may affect the design objectives, such as budget, energy consumption, environmental considerations, problems of access and so on. These constraints must be recognised at the objectives stage of the design.

A.1.4.3 Specification

Once the design objectives have been defined, they must be quantified wherever possible. However, not all design objectives can be quantified. For example, the need to make an environment appear efficient cannot be quantified. Furthermore, although many objectives can be expressed in physical terms, suitable design techniques may not exist or may be too cumbersome.

For example, shadows caused by pipes and containers on a pipe rack area are difficult to predict accurately unless three-dimensional imaging is used. This does not mean that objectives falling into this category should be ignored, but that experience and judgement should replace calculation.

A.1.4.4 General planning

When the design specification has been established, the remaining stages of the design should translate these requirements to the best possible solution. At the general planning stage, the designer aims to establish whether the original objectives are viable, and resolve what type of design can be employed to satisfy these objectives.

Assuming the original objectives are viable, the next choice for the designer would be that of lamp source. Lamp sources are many and varied, each having its own advantages and disadvantages. For example there is often a balance to be drawn between lamp efficacy and colour rendition. With this in mind, the tables in Section 3.2 may prove useful.

Having selected the lamp source, the next stage would be to match this source to the most suitable luminaire. This is perhaps the most difficult task due to the plethora of manufacturers and luminaire types available. It is obvious that the selected luminaire/source could possibly be improved upon if *all* the known variants were known. Since this is not feasible, the final decision will be based upon the designer's experience and knowledge of known manufacturer's products.

A.1.4.5 Detailed planning

Detailed planning (design) involves the calculation of such things as the number, position and aiming of luminaires, glare calculation, final costs and so on. Once the decision has been made as to the source and luminaire which is believed to be most suited to the area to be lit, calculations should be carried out to show the locations where the luminaires and source combinations will be effective. As only outline calculations can be carried out by hand due to time limitations, a computerised lighting design program is usually used. It is possible to link packages to Computer-Aided Design (CAD) drawings of the area to be illuminated.

A.1.4.6 Assessment of complete projects

When an installation is complete it is usual practice to carry out an inspection during the hours of darkness. Section 4.5.6 provides a useful checklist for this.

A.1.5   Road lighting design parameters

A.1.5.1   General

The ability of the eye to see in road lighting conditions is dependent on the following parameters:

- Average level of luminance in the field of view
- The uniformity of that luminance
- The quantity of glare present.

The average level of luminance $\bar{L}$ in the field of view is an influential parameter because it determines the state of adaptation of the eye. This state of adaptation in turn determines the efficiency with which the eye performs as measured by a number of indicators such as detection of contrast, sensitivity to glare and recovery time. The higher the $\bar{L}$, then higher the state of adaptation and better the eye performs. In the low lighting levels provided by road lighting, colour vision is poor and small dark objects are seen by the luminance difference (contrast) between them and the background. When the object is darker than the background the contrast is negative and this is described as 'silhouette vision'. The threshold contrast (C_{th}), that is the just perceivable contrast, decreases as the background luminance increases.[1] In other words the chances of being seen increase as the background luminance increases. Large and/or bright objects will be seen in positive contrast, that is direct vision, and many objects will be seen in a combination of negative and positive contrast.

Luminance uniformity (U_o) is a parameter because the contrast sensitivity of the eye is lower if there are large luminance differences in the field of view. The brighter areas can be considered to provide glare sources when viewing the darker areas.[2] The overall uniformity of the scene is therefore important.

Glare is a parameter because of the negative effect it can have on a person's visual performance. Glare is caused in the eye by unfocussed light falling on the focussed image of the target, producing an equivalent veiling luminance L_v which reduces visual performance. This can be represented at road lighting luminance levels and angles, by the empirical formula

$$L_v = 10\frac{E_{eye}}{\Theta^2}\, \mathrm{cd\,m}^{-2}$$

where E_{eye} is the illuminance on the eye produced by the glare source(s) in a plane perpendicular to the line of sight (lux); Θ is the angle between the direction of view and the direction of light incident from the glare source(s) (degrees); and 10 is a constant of proportionality in (degrees^2/steradian) which increases with age of the observer.[3]

An object at the threshold contrast can just be seen when there is no glare. It will not be seen in the presence of glare (unless the contrast is increased). The amount of extra contrast required to make the object just visible again under the glare conditions is called the threshold increment (TI), and in road lighting conditions this TI is taken as a measure of the glare level.

Longitudinal uniformity of luminance U_l, the ratio of minimum-to-maximum luminance along the centreline of any traffic lane, impacts only the visual comfort rather than visual performance. However, it should be noted that extensive exposure to low visual comfort may affect visual performance. In addition observers tend to have a poor opinion of roads with low longitudinal uniformity.

Surround ratio is applied because drivers need to see hazards on the roadway and also people animals and vehicles at the side of the road who may well become hazards. It is therefore useful to add this criterion giving a minimum limit to the ratio of average illuminance of a strip of land alongside the carriageway to the average illuminance of the adjacent strip on the carriageway.

A.1.5.2 Visual performance criteria

A.1.5.2.1 GENERAL

The driving task is a complex one which can be regarded as consisting of positional, situational and navigational sub-tasks and has been fully described in C.I.E. Report No. 100. The visual performance of the driver is a crucial element in identifying input to such tasks and has been assessed relative to the road lighting design parameters using the criteria of visibility distance, detection probability, revealing power, supra-threshold visibility, reaction performance, detection of relative movement and others. As the type of lighting that would improve one of these criteria might not enhance others, road lighting design emerges, as do many engineering designs, to be the search for the best available compromise carried out in terms of the various parameters.

A.1.5.2.2 VISIBILITY DISTANCE

Visibility distance has been determined[4] using 0.2 m × 0.2 m objects of known contrast placed upright on a road driven by observers at various speeds and wearing spectacles of various transmission factors to replicate

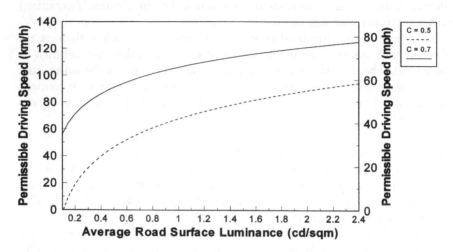

Figure A.3 Safe driving speed for average road surface luminance with different object contrasts [Economopoulos].

different lighting levels. Visibility distance has been found to increase with increase in background luminance. Then, knowing driver reaction time and vehicle deceleration rate, safe speed limits can be allocated to average road surface luminances as shown in Figure A.3. A typical result is that

- With an object contrast of 0.5 and an average road surface luminance of $2\,\mathrm{cd\,m^{-2}}$, $90\,\mathrm{km\,h^{-1}}$ is safe.

A.1.5.2.3 DETECTION PROBABILITY

Detection probability has been examined using a scaled simulator and apparent size objects again of $0.2\,\mathrm{m} \times 0.2\,\mathrm{m}$. This arrangement was used to investigate the relationship between the average road surface luminance and the overall uniformity required for a 75 per cent probability of detection of a small dark object in the darkest part of the road.[5] It was found that

- The probability (75 per cent) was obtained when the average road surface luminance was close to $1.5\,\mathrm{cd\,m^{-2}}$ and U_o was 0.4. Such figures are the basis for British Standard recommendations for non-motorway high speed and dual carriageway roads; and
- If the uniformity was halved, a fourfold increase in $\bar{L}$ was required to restore 75 per cent detection probability (Figure A.4).

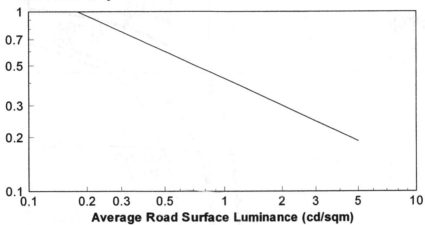

Figure A.4 Relationship between average road surface luminance and overall uniformity needed for a 75 per cent probability of detection of a 0.2 m × 0.2 m object in the darkest part of the road [Narisada].

A.1.5.2.4   REVEALING POWER

Revealing power is the percentage of objects detectable at a point in the field of view and is calculated from the distribution of illumination on the vertical faces of objects in the road and the statistical distribution of reflection factors of such objects.[6] Van Bommel in 1979 extended the work of Waldram, reaching, inter alia, the following conclusions (as shown in Figure A.5):

- An increase in $\bar{L}$ leads to an increase in revealing power, at least until $\bar{L}$ has reached nearly 10 cd m^{-2}
- Increasing the TI decreases the revealing power, for example from 85 to 70 per cent when TI changes from 7 to 30 per cent with $\bar{L}$ remaining constant
- A halving of overall uniformity from 0.4 to 0.2 at $\bar{L} = 1.1$ cd m^{-2} requires a 3.5-times increase in $\bar{L}$ to maintain 75 per cent revealing power.

The first finding confirms the importance of the average luminance of the road surface. The second shows the detrimental effect of glare. The third agrees with the four-times factor found for Narisada's detection probability.

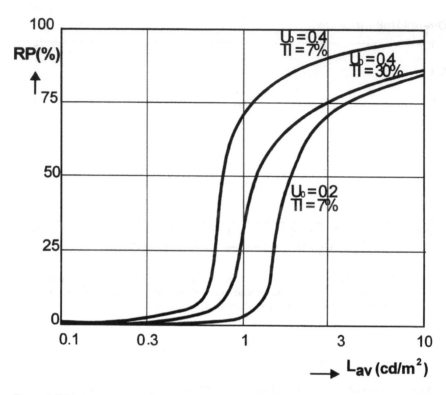

Figure A.5 Variation in revealing power with average road surface luminance [van Bommell].

Note: A revealing power of 75 per cent does not imply that 25 per cent of objects cannot be detected. The concept is based upon small objects and larger objects can be considered as a number of smaller ones. For example a pedestrian may be considered as being perhaps four such objects. In addition larger objects are likely to cross boundaries between regions of different revealing power. Also consider that the calculations are for an instantaneous element of a dynamic process, with the object being seen against a changing background as it is approached.

A.1.5.2.5 VISIBILITY INDEX

Visibility index (VI) is used in America and is simply the visibility above the threshold level[7] where

$$VI = \frac{C}{C_{th}} DGF$$

where C is the task contrast under the road lighting conditions actually prevailing; C_{th} is the threshold contrast of the CIE reference task at the background luminance L_b; and DGF is the disability glare factor which itself is given by

$$DGF = \left(\frac{L_b}{L_{bg}}\right)\left(\frac{C_{th}}{C_{thg}}\right)$$

where L_{bg} is the effective background luminance with glare present; and C_{thg} is the threshold contrast at L_{bg}.

An advantage of VI is that it can be calculated from normal lighting parameters. Following calculation the effect of the parameters $\bar{L}$, U_o and TI on it can be examined. Van Bommel in 1979 found that

- An increase in $\bar{L}$ up to about $10\,cd\,m^{-2}$ leads to an increase in VI
- Increasing the TI from 7 to 30 per cent with U_o constant requires a 2.6-times increase in $\bar{L}$ to maintain VI at 6, which is roughly equivalent to 75 per cent revealing power
- A halving of overall uniformity from 0.4 to 0.2 at $\bar{L} = 1.1\,cd\,m^{-2}$ requires a 2.4-times increase in $\bar{L}$ to maintain VI at 6. The latter finding compares with the 3.5-times derived from Waldram's revealing power and with Narisada's 4-times for detection probability.

A.1.5.2.6   DETECTION OF RELATIVE MOVEMENT

Detection of relative movement is linked to the change of angular size of the rear of a vehicle and has been studied by laboratory simulation. Fisher and Hall in 1976 found that

- Detection time decreased rapidly as $\bar{L}$ increased
- A higher $\bar{L}$ was required to maintain detection time for slower lead car decelerations
- For increases in $\bar{L}$ above $10\,cd\,m^{-2}$ little further benefit accrues.

A.1.5.2.7   SUMMARY OF VISUAL PERFORMANCE PARAMETERS

The visual performance parameter research can be summarised thus:

- Visual performance increases quickly as background luminance is increased up to between 1 and $2\,cd\,m^{-2}$ and more slowly until it levels out at as the luminance approaches $10\,cd\,m^{-2}$
- If the overall uniformity U_o is reduced then visual performance is reduced and such a reduction can be compensated for by an increase in background luminance
- An increase in glare reduces visual performance and such a reduction can be compensated for by an increase in background luminance.

A.2 Equipment

A.2.1 Corrosion

Corrosion in metals usually requires a combination of moisture and oxygen. Protecting the raw metal from both, by for example painting or galvanizing, is an effective defensive approach. Damage that may potentially breach the protection should be considered at the design stage, for example handling (causing scratched paint) and in-service (erosion of protection by weathering). Certain metals and metal alloys naturally resist corrosion. Stainless steel forms a chrome oxide surface layer that rapidly reforms when damaged and protects the underlying material (depending on the grade used). Aluminium also forms an oxide-protective layer with some degree of protection. Copper alloys such as brass and bronze have degrees of corrosion resistance: bronze has been widely used in marine applications.

Dissimilar metals placed in contact in the presence of moisture set up an electrolytic action that may accelerate corrosion of one of the metals. (Further information is given in PD6484.)

Typical examples of electropotential values of various metals and metal alloys are shown in Table A.1. If the difference in voltage between two metals in contact is large enough significant corrosion will occur dependent on the environment and the nature of the electrolyte. The maximum allowable potential differences are shown in Table A.2.

The effect can be overcome by a suitable choice of compatible materials or by separating incompatible materials (e.g. by plastic washers, suitable paint films or non-conductive surface films).

A.2.2 Ingress and mechanical protection

Ingress protection relates to the ability of an enclosure (including a luminaire) to withstand the ingress of both solid objects and dust (dirt), and

Table A.1 Electropotential values of various metals and metal alloys

Metal/metal alloy	Electropotential (V)
Aluminium	−0.75
Brass	−0.3
Cast iron	−0.7
Copper	−0.18
Gunmetal	−0.24
Steel	−0.58 to −0.79
Stainless steels	−0.2 to −0.45
Zinc	−1.1

Table A.2 Maximum allowable potential difference to avoid corrosion by
          electrolytic action

Environment	Maximum allowable potential difference
Liable to contamination by sea water; normally exposed to the weather	limit 0.25 V
Interior part exposed to condensation but not to salt	limit 0.5 V
Interior parts that remain dry	greater than 0.5 V

Note
A.2 The metal with the more negative value will tend to corrode.

moisture. A numbering system classifies the degrees of ingress protection as IPXX; the first numeral relates to solid objects and dust, the second to moisture. The two numerals are not directly related to each other. Degrees of protection appropriate to luminaires are shown in Table A.3.

Table A.3 Ingress protection against solid objects and dust, and moisture

Protection against	IP rating	Description
Solid objects	IP2X	Protected against a 12 mm (test) finger (safety against touching live parts)
	IP3X	Protected against a 2.5 mm probe (safety against touching live parts, e.g. screwdriver)
	IP4X	Protected against a 1.0 mm probe (safety against touching live parts, e.g. by a wire)
Dust	IP5X	Minimal powder ingress such that it would not cause an electrical fault if conductive
	IP6X	No powder ingress at all
Moisture	IPX2	Drip proof (e.g. condensation falling within a lighting column onto the backboard circuit)
	IPX3	Rain proof (represented by water sprayed at $\pm 60°$ from nozzles on a semicircular hoop)
	IPX4	Splash proof (represented by water sprayed from every direction from nozzles on a semicircular hoop)
	IPX5	Jet-proof (water sprayed to represent e.g. hose cleaning)
	IPX6	Powerful water jet-proof (water sprayed at high pressure and volume to represent, for example, heavy seas or pressure cleaning)
	IPX7	Watertight (occasionally covered in water but not for operation under water)
	IPX8	Pressure watertight (for operation under water)

The dust tests indicate the protection against degradation due to dirt entering the luminaire and reducing light output in both quantity and distribution. Maintenance factors in the lighting design standards take account of this. A high IPX is of value since cleaning the inside of a luminaire effectively is usually difficult and relatively expensive.

A typical dust test apparatus is shown in Figure A.6. The chamber housing the operating luminaire produces a cloud of talcum powder of a standard size which surrounds the luminaire. The luminaire is switched off and the test runs for 3 hours. During this time the luminaire will cool and try to draw powder in past the seals.

Exterior luminaires are often required to withstand hosed water, IPX5 or X6. The test requires the energised luminaire (hot) to be hosed for 3 min from all directions when it has been switched off. The partial vacuum created tries to draw water across any weakness in the seals.

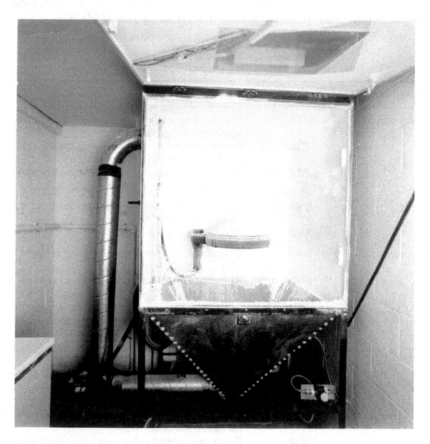

Figure A.6 Typical dust test apparatus.

Figure A.7 Typical moisture test apparatus.

A typical test apparatus for IP moisture tests is shown in Figure A.7. For luminaires with drain holes, water entry including condensation is allowed during tests if it can drain out effectively. No water ingress is permitted for luminaires without drain holes. No contact is permitted through the drain holes with live parts with the relevant test probe for IP3X and IP4X luminaires.

IKXX values are also used in the index of protection. This figure refers to the degree of protection the appliance offers against bumps and knocks and mechanical impacts. The protection offered is described in terms of impact energy measured in joules, as delivered via a pendulum device. Table A.4 shows the indices that apply to the different impact energies delivered by the pendulum.

Table A.4 Protection against mechanical impact

Code IK	IK01	IK02	IK03	IK04	IK05	IK06	IK07	IK08	IK09	IK10
Impact energy (joules)	0.15	0.20	0.35	0.50	0.70	1	2	5	10	20

A.2.3 Characteristics of materials for use in luminaires

A.2.3.1 Housings – metals

A.2.3.1.1 ALUMINIUM

In various forms aluminium is a commonly used material. It can be relatively inexpensive, is lightweight and strong and consequently is used in a wide variety of luminaires. Key characteristics are

- It is readily formed, by casting in sand, gravity or pressure die, or by spinning, pressing, extrusion or fabrication
- Appropriate alloys give natural corrosion resistance or can be readily treated to protect them, for example silicone-rich casting alloys withstand marine conditions even in a natural finish
- It takes decorative and protective finishes well
- It is available with high light reflectivity and ability to be finished up to a specular polish, allowing the housing to be used as the reflector
- Joining is by welding, adhesives or mechanical fasteners
- It is recyclable.

A.2.3.1.2 STAINLESS STEELS

Stainless steels can be relatively expensive, heavy and strong. A key attribute is its natural resistance to corrosion. It also has an attractive natural appearance when mechanically finished by either brushing or polishing. It is generally used where appearance and high corrosion resistance are required. Key characteristics are

- It is fairly hard and therefore more difficult to form
- It is usually fabricated from sheet, or when for parts of housing machined from stock or spun
- Spinning requires fairly expensive annealing as the material quickly work-hardens
- It can be painted but paint adhesion can be a problem unless the correct pre-treatment is applied
- Although a polished specular finish is possible it has a fairly poor reflectivity (0.6 or so) so is not a good optical reflector
- Joining is by welding, adhesives or mechanical fasteners.

A.2.3.1.3 COPPER

Copper is relatively expensive and reasonably strong. It is corrosion resistant in its natural form, and is usually used for both its attractive appearance and ability to be joined by solder for strong clean joints without witness marks, for example, on thin structures where appearance is paramount, such as lightweight frames. It is widely used for heritage luminaires. Its key characteristics are

- It is readily formed by spinning, pressing and fabrication
- It takes decorative finishes well, provided the correct pre-treatment is applied
- It polishes well to give an attractive appearance that can be protected by clear lacquer
- Unprotected the finish weathers to a pleasant coppery brown
- Joining is usually by soldering and mechanical fasteners
- It is recyclable.

A.2.3.1.4 COPPER ALLOYS

Copper alloys such as brass and bronze are relatively expensive, heavy and strong. They are used for their natural attractive appearance and corrosion resistance. Typical applications would be in marine situations and decorative luminaires where the natural colour is valued. Key characteristics are

- They are readily formed by casting (in by sand or gravity), spinning, pressing, extrusion and fabrication
- They take decorative and protective finishes well, or alternatively, weather to attractive dull colours
- They can also be polished and protected by clear lacquer
- Joining is by brazing, solder, welding, adhesives or mechanical fasteners
- They are recyclable.

A.2.3.1.5 STEEL

Steel is low cost, heavy but strong even in thin sections. However, it requires reliable protection against corrosion. It has a wide range of uses, for example fabricated bodies, control gear boxes. Key characteristics are

- It is readily formed by spinning, pressing or fabrication
- Protection can be by paint alone, and it takes decorative finishes well; however, paint finishes are susceptible to damage and subsequent rapid rusting
- Electroplating (e.g. zinc) improves resistance to corrosion but is a thin protection and is also easily damaged
- Thermal metal spray, usually zinc or aluminium, is better than electroplating; however, the most effective treatment is hot-dip galvanising, either after fabrication or by use of pre-treated steel sheet
- Joining is by welding, adhesives or mechanical fasteners.

A.2.3.2 Housings – plastics

A.2.3.2.1 GENERAL

A wide range of plastics are used for housings often as a major part allied with metal components to add strength. The advantages of many plastics are

- They are easily formed into complex shapes by moulding (thermoforming, or injection and compression moulding)
- Compression and injection moulding allow accurate details to be formed, such as fixings or locations for components; this minimises assembly cost and time albeit with fairly high tool cost
- A wide range of colours can be integrated within the material to save finishing
- They can take paint finishes and printed detail, and textures can be moulded in
- They are relatively inexpensive, lightweight and fairly strong.

Disadvantages are

- Many plastics are prone to degradation, for example from UV radiation that embrittles and discolours, and natural weathering that erodes the surface; surface treatment (e.g. painting) and chemical inhibitors (e.g. UV absorbers) can delay the process
- Some plastics, particularly thermoplastics (that are shaped by heat but whose structure does not change) are prone to creep under load, that is they deform
- Thermoplastics also have limited heat resistance, and must be chosen to suit the likely thermal conditions they will experience, for example from lamps and control gear
- Some plastics absorb water and deform
- Recycling is limited.

However, the many advantages of plastics have enabled their use to significantly reduce product costs, with their disadvantages being mitigated by design and processing techniques. For example

- For degradation, UV absorbing additives will delay the UV attack on the plastic, and paint finishes will form a barrier, and eventually degrade themselves and
- For strength, additives to the raw plastic such as glass fibres can add considerable strength, and metal components can be moulded in to reinforce structural areas.

A.2.3.2.2 ABS (ACRYLONITRILE-BUTADIENE-STYRENE)

ABS is a low-to-medium-cost thermoplastic. Its characteristics are

- It is easy to mould or thermoform
- It has medium strength
- It is used in opaque form for covers of housings
- For exterior use it requires protection from UV, by blending or painting
- It has a maximum temperature around 95 °C.

A.2.3.2.3 POLYCARBONATE

Polycarbonate is a medium-to-high-cost thermoplastic. Its characteristics are

- It is fairly easy to mould or thermoform
- It has high strength, and additional strength can be provided from incorporation of glass fibre
- It is used in clear and opaque form for housings
- For exterior use it requires protection from UV, by additive absorber, surface film or painting
- It has a maximum temperature of around 130 °C.

A.2.3.2.4 POLYAMIDE (NYLON)

Polyamide is a medium-cost thermoplastic. Its characteristics are

- It is fairly easy to mould
- It has high strength, and additional strength can be provided from incorporation of glass fibre
- It is used in opaque form for housings
- For exterior use it requires protection from UV, by additive absorber, surface film or painting
- It has a maximum temperature of around 120 °C.

A.2.3.2.5 GRP (GLASS-REINFORCED POLYESTER)

GRP is a low-cost thermoset plastic, a blend of resin, mineral filler and chopped (DMC) or stranded (SMC) glass fibre. Its characteristics are

- It can easily be compression moulded
- It is a medium-strength engineering material, the strength derives from the fibres
- When used for housings, low expansion allows metal reinforcing inserts to be moulded in

- It has minimal creep
- As it has a low plastic content it has good resistance to UV
- Gradual erosion of surface exposes fibres, which can lead to problems of skin irritation during maintenance, and also to a loss of gloss and colour, although it can be painted initially or in service
- It has a maximum temperature of around 130 °C.

A.2.3.3 Glazing – glass

Glass has many advantages in terms of great durability in a wide range of environments, ability to be moulded into prisms (by pressing), a low-cost base and it can be cleaned effectively. However, it has the disadvantages of weight, and vulnerability to damage. It is available in clear, limited patterns, and opal. It is widely used in flat form as covers for floodlights and glazing to road lights. A thermal-toughening process is essential to give resistance to thermal shock, for example rain falling on to the hot glass of a luminaire, and improved impact resistance. In toughened form its ability to operate at up to 250 °C without degradation is unsurpassed. It retains clarity without any UV degradation. Moulding and blowing produces shaped glazing bowls. Prisms for light control are readily moulded, although not as sharply as with thermoplastics so there is less precise light control. Shaped glass is less easy to toughen, and prismatic forms cannot be thermally toughened, as the process begins to re-melt the prisms. Low expansion glass formulations such as borosilicate are used to give these items resistance to thermal shock.

A.2.3.4 Glazing – plastics

A.2.3.4.1 ACRYLIC (PMMA – POLYMETHYLMETHACRYLATE)

Acrylic is a very durable thermoplastic that is available as flat sheet that can be thermoformed, and, granules for extrusion of linear prisms, or injection moulding of complex shapes and prisms. It is available in clear, limited sheet patterns, colours and opal. It is not very susceptible to UV and typically can last 10 years, with good retention of clarity but gradual susceptibility to crazing. It is not impact resistant. Grades of toughened acrylic are available as moulding granules with rubber additives to improve this, but can lead to a change in clarity giving a slight milky appearance when warm. It is widely used in luminaires, and has a maximum working temperature of 85–100 °C depending on the grade.

A.2.3.4.2 POLYCARBONATE

Polycarbonate is a fairly durable thermoplastic that is available as flat sheet that can be thermoformed, and granules for extrusion of linear prisms, or injection moulding of complex shapes and prisms. It is available in clear, limited sheet patterns, and opal. It is more susceptible to UV than acrylic, with a typical life of 5–10 years, but with gradual yellowing and loss of strength. Surface films and/or UV additives improve its resistance to UV. Careful luminaire design regarding temperature of the material in service is essential to maximise durability. Its great advantages are its impact resistance, it can take several years to reduce to that of acrylic, and its working temperature of around 130 °C. It is widely used in luminaires, particularly where vandalism is an issue. The maximum working temperature reduces to 100 °C where UV is also present, for example when used as glazing. Above this temperature degradation occurs more rapidly.

A.2.4   Luminaire light performance measurement

The distribution of intensity in all directions from the luminaire is measured at intervals sufficiently close together that will allow an accurate interpolation of intensity in any direction, including those not directly measured. Measuring point spacings are recommended, for example, by the CIE. The resultant table of intensities and angles is known as an *I* table and when produced in standard formats (e.g. TM14) can be used by installation design software to produce a wide range of lighting data for both the individual luminaire and the installation it is used in.

Accurate photometric data requires suitable equipment with appropriate measuring practices. Lamps should be selected as representative of their specification and aged to give a stable light output. Since their light output is sensitive to supply voltage this needs to be stabilised; some lamps are temperature sensitive and so the measuring environment needs to be controlled, normally to 25 ±2 °C. The measuring cell must be stable and have a linear response (to record accurately over a wide intensity range) and the measuring distance from the cell to the luminaire must be at least 5 times the maximum optical dimension of the luminaire (to minimise inverse square law inaccuracy). For a luminaire 1200 mm × 600 mm (e.g. a fluorescent module) this requires a light path of nearly 7 m. The cell must only see light from the luminaire in the direction required, that is it must be shielded from other light, including reflected.

A typical photometer is shown in Figure A.8. It is housed in a space with all surfaces blackened to minimise reflected light. The light path is via two mirrors to compress the 7 m into a physically smaller space. The photocell is shielded so that it only sees the luminaire as reflected by the mirrors. The mirror arm rotates vertically around the luminaire to allow the vertical distribution of light to be measured in a given plane. The luminaire is then rotated about its vertical axis to allow a new vertical plane to be measured, and this is repeated until the whole light distribution has been recorded. The *I* table is recorded on a computer and issued electronically for processing.

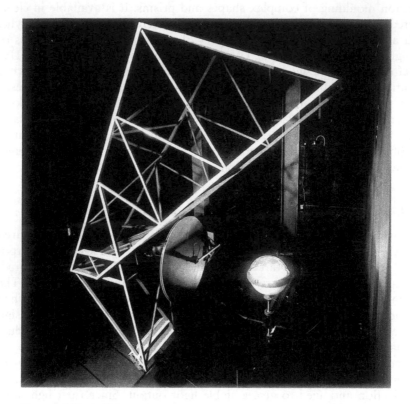

Figure A.8 Typical photometer.

A typical luminaire emitting light downwards only will have some 1300 separate intensity measurements made.

A.2.5 Floodlight beam types

Floodlights are generally categorised into three main types by reference to the beam produced: The three beam patterns, (a) symmetrical, (b) asymmetrical and (c) double asymmetrical as shown in Figure A.9.

A.3 Particular applications

A.3.1 The outdoor environment

A.3.1.1 General

The following are areas of potential difficulty that might be encountered (by the inexperienced) when lighting the outdoor environment.

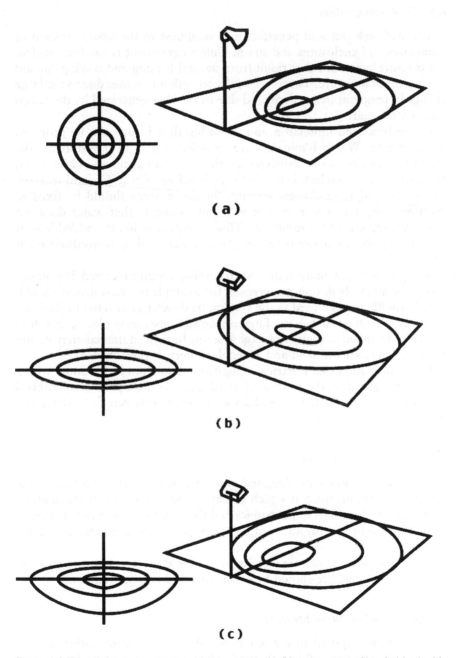

Figure A.9 Floodlight beam types: (a) symmetrical, (b) asymmetrical and (c) double asymmetrical.

A.3.1.2 Weather: Rain

Water will seek out and penetrate any weakness in the sealing system of luminaires and enclosures and so on. Unless equipment is carefully sealed, the pressure variations that result from natural heating and cooling can aid the ingress of moisture. This is especially so with luminaires due to the large change in temperatures experienced due to the heat generated by the lamps and control gear.

Ground-recessed luminaires and those installed in pits can be prone to water ingress. Where luminaires are installed in pits in paved areas, the pit and luminaire can be protected by the use of an armoured glass cover; however, care should be taken to limit the build-up of heat. Ground-recessed luminaires and LED clusters installed in paved areas should be fixed in position using tile spacers and resin grout to ensure that water does not run down the side of the luminaire. This arrangement has the added benefit of retaining the luminaire firmly in place. Silicon sealing is insufficient for either purpose.

The ground area immediately surrounding ground-recessed luminaires should be a readily draining aggregate, for example pea gravel, and should, together with any pits used in the system, be directly connected to the local drainage system or soakaway. Engineering sites are messy places, therefore it is recommended that any drainage systems be proven immediately before the luminaires are installed and at regular intervals thereafter.

Luminaires and other electrical equipment should not be mounted with a cable-entry gland on the top side if at all possible. Ballast or other control gear that are not coated to withstand water ingress can fail even when within gear-enclosed luminaires

A.3.1.3 Weather: Temperature

All equipment should be designed and specified to take account of the extremes of temperature at which it may be operated. This is particularly important for locations in the tropics and the polar regions where temperatures can be substantially above or below those experienced in the temperate zones.

As the temperature of bare metalwork can be at extremes, access arrangements for maintenance should be easily operated by a gloved hand.

A.3.1.4 Weather: Wind pressure

All equipment exposed to the wind should be designed to withstand any loadings to which it will be subjected. This is particularly true for lighting columns, high masts and other structures that raise the height of equipment. Care should be taken to ensure that all attachments that increase the loading on the equipment are taken into account in the design.

Such attachments include

- Signage
- Bunting
- Banners
- Flower baskets
- Supplementary luminaires
- Festive decorations.

If lighting columns or other supports do not have adequate strength they will suffer fatigue, increased distortion of the material and accelerated corrosion at areas under stress. Under extreme conditions lighting columns or supports could collapse in service.

Care should be taken to ensure that all lighting columns are installed to the manufacturers' recommendations. Planted root lighting columns should be erected and the excavated hole refilled with sufficient material of the correct type to ensure that the loads imposed above ground are counteracted by the foundation. This may require the use of concrete as against the excavated material. Flange-plate–mounted lighting columns should be installed on a purpose-designed foundation to suit the particular ground conditions and the loads imposed. Such foundations should be designed and specified by a suitably qualified structural engineer.

A 10-m column with a 1.5-m planting depth will have a lower probability of remaining vertical if planted 300 mm shallow, as it will then be a 10.3-m column with a planting depth of only 1.2 m.

Foundations for columns planted near the top of embankments or in soft soil or sand should be carefully considered as their roots may not be subject to the soil pressure of those planted elsewhere or in heavy soil conditions. In such cases the use of a specially designed concrete foundation may be required and the advice of a structural engineer should be sought.

Wall-mounted equipment is also exposed to the effects of wind pressure which may be increased by localised turbulence and gusts created by the surrounding structures. Care should be taken when installing wall-mounted equipment or anchorages for cross-road spans and so on to ensure that wall fixings are suitable for the loads being imposed and that the load is spread over a sufficiently large area of the wall or structure. Multiple anchorages or wall fixings should not be installed in the same brick or piece of masonry if at all possible. Any protective system or decorative coatings on walls or structures should not be damaged during installation.

A.3.1.5 Weather: Atmospheric condition

All exterior lighting will be affected to a greater or lesser extent by weather conditions. Rain and mist will increase the refractions of light causing a general glow to form over and around the luminaires, and, as the density of the rain or mist increases, over the lit area as a whole.

Thick fog and blizzard conditions will render any exterior lighting installation ineffective due to the high level of refraction and subsequent light scatter from the water molecules and ice particles in the atmosphere.

The refraction of light passing through the air is not only increased by the density of the water content and solid particulates in the atmosphere but also by the length of the light path. Therefore, the effect can be reduced by shortening the length of the light path. This can be achieved by using lower mounting heights and/or lower luminaire elevation. However, this may seriously affect the cost of the installation and its subsequent maintenance costs by increasing the number of luminaires required and their supports. Such remedies should only be considered where there is a high requirement to provide lighting under all conditions.

A.3.1.6 Corrosion resistance

Due to the presence in the outdoor environment of water, wind, sea spray, vehicle emissions and corrosive soil types, corrosion protection of equipment has a far greater importance than indoors.

Protective systems that use plating, metal spraying or other coating techniques cannot be relied on unless all surfaces including the inside of enclosures, lighting columns and other equipment exposed to the weather are adequately protected. The maximum electropotential difference of metals in contact should not exceed 0.25 V externally (see Section A.2.1). All protective system need to be adequately and regularly maintained if they are to protect the equipment and maximise its in-service life. All mild steel items must be hot-dip galvanised as a minimum.

A.3.1.7 Vandalism

Experienced lighting practitioners know that the problem of vandalism has neither diminished nor increased in recent years. In many places it lies dormant waiting for them to drop their guard. The increasing emphasis being placed on a combination of daytime aesthetics and interest and diversity at night can produce the opportunity for vandalism to thrive unless careful consideration is given to it at the design stage. It is often considered that equipment in town and city centres will not be subject to vandalism due to the increased use of CCTV surveillance and so on. However, this has been proven to be erroneous with expensive equipment being badly damaged by vandals.

Ground-recessed luminaires and those installed in pits are particularly prone to vandalism and should only be used where there is a high level of pedestrian movement and other observation or they can be situated in areas away from pedestrian and vehicular movements. Hard sharp objects can easily shatter visor glass that can withstand a lorry wheel. Once such luminaire visors are damaged, there are immediate issues with regard to laceration and exposure to live electrical contacts. This is a particular problem in areas where young children may explore the damage before it is repaired or made safe by maintenance personnel. If there is rainfall after the damage and before repair, then components within the luminaire will require to be dried or they are likely to fail. Where possible, pits should be covered by a metal grill to minimise damage by vandals.

Vulnerable equipment such as ground-recessed luminaires or those installed in pits should be fed from a separately fused and protected power circuit to reduce the possibility of leaving a whole area without lighting.

For vandal-prone areas, some manufacturers can provide a discreet wire mesh around the visor; however, this will not resist air-gun pellets. In such areas lighting column doors can be installed higher up or in extreme cases near the top of the column, however, consideration must be given to access for maintenance and damage to electricity supply cables in the event of a vehicular impact.

Where there is no exposed visor, for example some indirect luminaires, then fewer problems are experienced from vandalism. All luminaires in pedestrian underpasses should be installed with vandal-resistant visors and/or a protective metal grill.

Equipment that is primarily designed for the indoor environment is unlikely to be suitable for outdoors. For example wall-recessed luminaires that clip into place can therefore be easily levered from their fixing.

A.3.1.8 Equipment design

The maximum surface temperature of visors or metal work of ground-recessed luminaires that are installed in areas accessible to the public must be less than 70 °C at an ambient temperature of 30 °C. However, at this temperature blistering of the skin may still occur and it is recommended that the temperature of the visor and associated metal work should not exceed 65 °C at an ambient temperature of 30 °C. For areas near swimming pools, paddling pools or other wet areas, the temperature should be limited to 40 °C.

A.3.1.9 Glare

The effects of glare are widely misunderstood and commonly ignored in external lighting installations. At best, glare results in an annoyance to the observer and at worst a loss of visual performance. The best known and most easily implemented guide is that there will be unacceptable glare present if an observer at his normal viewing angle can see the lamp in a luminaire (rather than the optic). There is no reason glare should be present. It can be removed by intelligent design and the use of louvres, cowls and shields, and so on (see Section 2.3).

One of the most common sources of glare is from luminaires mounted with too high an elevation angle (see Figure 2.1). This usually means that the luminaires are not mounted sufficiently high, the wrong luminaire has been selected or both. Another source of glare is from area illumination using wall-mounted luminaires with refractor visors mounted vertically. These should be avoided in favour of luminaires that emit light in a downwards direction.

A.3.1.10 Light pollution

Light pollution is closely linked to glare in that many of the causes and solutions are similar.

The International Dark-Sky association asks the simple question, 'Why would anyone wish to pollute the night sky?' and comes to the conclusion that much poor lighting is caused by users being unaware of the issues and what is good lighting practice.

Poorly aimed floodlights are a common source of light pollution. To light a horizontal plane (i.e. the ground) there is no reason that the light-emitting face of luminaires should be angled at more than 20° from the horizontal, and preferably the face should be parallel to the ground. If there is difficulty achieving this then the luminaire mounting height should be raised to allow a reduction in the angle. Alternatively additional and/or different luminaires should be installed. A lighting scheme that does not achieve these guidelines will project unwanted light into the sky, and as a consequence waste money and pollute the environment.

Luminaires mounted lower than the object to be lit inevitably produce some light directly into the sky and therefore cause light pollution. Good application of the underlying scientific principles together with good luminaire selection can minimise the effect.

However, to minimise pollution objects should be lit from above wherever possible. The trend to minimise light pollution by use of flat-glass visors has produced schemes that do not appeal to untrained observers, giving a feeling of being lifeless and without 'sparkle'. For the best result there has to be a balance of low light pollution and some 'sparkle' to the installation. A small amount of light emitted at higher angles can help to remedy this situation. Consequently the use of low profile bowls (<60 mm) rather than flat-glass visors is recommended for areas with high night-time pedestrian movement and high amenity. However, flat-glass luminaries are recommended for sensitive areas and rural traffic areas.

A.3.1.11   Aesthetics and visual intrusion

Exterior lighting systems can be visually intrusive both by day and by night. During the day the physical appearance of the installation may be seen from a considerable distance, particularly if high lighting columns or other supports are used. Where possible, lighting columns should be kept in proportion with the surrounding buildings and should not project above them. If possible, lighting columns should be viewed against a backdrop of buildings or trees where they will be disguised or hidden. This is not always possible, and in many instances the lighting columns will be seen against the sky. In these cases, the visual effect of the lighting columns can be reduced by careful choice of equipment and colour, that is use light neutral colours such as grey. Alternatively, a lower mounting height lighting column may be used. However, this may increase the number of lighting columns required thus increasing the visual clutter. This is a trade-off which must be carefully assessed for each individual situation.

There are few, if any, lighting solutions that require lighting columns to be spaced closely together. An absolute minimum separation between columns of 14 m in any direction is often used as a guide, and this dimension can generally be increased with higher mounting heights and the correct choice of luminaire.

Visual intrusion at night can vary from intrusive light entering a person's property to the irritation of a bright light seen against a jet black sky. If the recommendations given above for the control of light pollution are carefully followed then intrusive light into adjacent properties will be minimised thereby the irritation factor will be reduced. However, it may be impossible to fully reduce the view of a lit luminaire against a dark sky and further mitigation will only be achieved by reducing or limiting the hours of operation. This may be practical in some instances but in others where a 24-hour process is involved or security is paramount further compromise may not be possible.

When planning a lighting system in close proximity to an airport particularly near the end of the runway, consideration must be given to the height of the equipment. All airports have a defined obstacle limitation area surrounding them where the height of structures is strictly limited. The exact shape and extent of this area varies with the type and use of the airport and its topography. It can extend from 2 to 15 km from the centre line or end of the runway, however, it is normally only necessary to evaluate the situation for any obstacles over 3 m in height within 3 km of the centre line or end of the runway. Full details of the restrictions and the area can be obtained from the local planning officer.

A.3.1.12 Electrical design

In principle the design of the electrical system for an exterior lighting system is no different to that for an indoor lighting system. However, the size of the area to be lit and limitations on available power supplies can give rise to added complications with volt drop and cable sizes. It is not unusual for cable runs of up to a kilometre to be installed to feed an exterior lighting system. This can result in cable sizes many times greater than those commonly required in interior lighting systems and thus associated problems with terminations, fusing and earthing. All of these problems can be overcome with sensible design and specification.

A.3.1.13 Vehicles

Vehicles, particularly the larger types, tend to show little respect for lighting equipment leading to the recommendation that lighting columns should be sited a minimum of 0.8 m and preferably 1.5 m away from kerbs or carriageway edge (see Table 4.26). If this is not possible or there are no kerbs, measures specifically designed to minimise the risk of damage by vehicles should be installed. This particularly applies to car parks where although vehicles are moving at low speed they necessarily approach objects from unusual directions and backwards.

Illuminated bollards should not be used as the primary barrier to prevent car parking on the footway.

It is a requirement of the Electricity Safety, Quality and Continuity Regulations that all overhead lines not exceeding 30 000 V mounted above a road accessible to vehicular traffic should have a minimum clearance from ground level of 5.8 m. In all other instances the minimum clearance is 5.2 m above ground level. For the avoidance of doubt it is considered that overhead lines should include all electrical equipment including luminaires.

A.3.1.14 Reflection from surfaces

Generally exterior lighting is more difficult to design than interior lighting because there are no ceilings and few walls surrounding the lit area. Those surfaces that do exist tend to have lower reflectivity than equivalent surfaces indoors. The contribution made by reflected light to the overall scheme is therefore lower with much more reliance being placed on direct illumination. As observers tend to be looking generally downwards, disability glare from luminaires having significant luminance (brightness) can be reduced by positioning them sufficiently far from the observer's field of view, usually higher. Inter alia, this means that bollards cannot be used to provide the primary light source where other than route identification is required as there will be little or no illumination on faces and objects above the height of the bollard unless excessive glare and light pollution is accepted. Similarly ground-recessed luminaires cannot adequately provide the primary lighting to a horizontal plane (i.e. the ground) due to the limited distribution of light which will not be incident to that surface.

The lack of walls surrounding exterior lighting installations can give rise to problems from not only the equipment and the lit area being on view but also the spread of the light outside the lit area. Careful design, specification and consideration of the problems as discussed earlier can help to mitigate these issues although the lit area will always be in view unless specific actions are taken to screen it by the provision of dense planting or screens.

A.3.1.15   Lighting system users

In many exterior lighting systems, the user in common with lit interiors is in a fixed location or only able to move around within a limited area relative to the lit area. In such situations the designer of the installation can optimise the design to suit the requirements and location of the user. A good example of this approach is the asymmetric (one-sided) additional lighting often provided for colour television purposes at football stadia. This lighting which is supplementary to the normal lighting system may only be provided from one viewpoint that of the television cameras. Elsewhere in the stadia the levels of illumination remain the same as those required for non-television use. The supplementary lighting can increase the vertical illumination in this one direction by up to four times.

A common problem found in many exterior lighting systems is that the user is mobile and not only able to move around the lit area but may do so at different speeds from normally walking pace to high speeds using mechanical propulsion. This increases the complexity of the design by requiring a higher uniformity of illumination and careful assessment of the glare from a number of directions. The problem may be further complicated by the user being at different heights within the lit area; for instance an operative in a container depot at ground level as against the operator of a crane in the same depot at 20 m above ground level. Both of these users will need different but complementary parameters of the lighting systems to be provided to allow them to carry out their respective tasks in safety and without strain. The increased speed of the crane operator may require a higher level of illumination than the person on the ground. However, the person on the ground may be more affected by the glare from the lighting installation due to having to look upwards whereas the crane driver will generally be looking downwards on to the scene.

A.3.1.16   Lighting system observers

Anyone observing a lighting system from outside the lit area may also be presented with similar problems to the system user. Again, careful consideration of these points needs to be done at the design stage if the lighting scheme is to have minimum effect on its surroundings. Particular attention should be given to the adverse effect the lighting may have on the safety and operation of nearby and adjacent transport infrastructure such as main roads, railways, canals and navigable waterways. Particular care should be given to ensure that lighting does not obliterate or confuse signalling systems on railways or destroy the night-time adaptation of the driver.

Aircraft landing at night use a system of approach and runway lighting which forms a specific pattern when viewed from the air. Care should be exercised to ensure that any lighting system illuminating a long narrow area such as a road do not conflict with or offer an alternative approach system of lighting on which a pilot may line up. This is a particular imperative when the area being lit lines up at a similar compass heading. This problem can be mitigated by the use of full cut-off luminaires as commonly used close to airports. However, as pilots need to recognise and line up on the approach lighting up to 16 km away from the end of the runway, the Civil Aviation Authority require all roads within 16 km of the airport to be considered in this manner. Lighting practitioners in doubt should contact the air traffic controllers of the specific airport for further help and advice.

A.3.2   Template schemes for popular sports

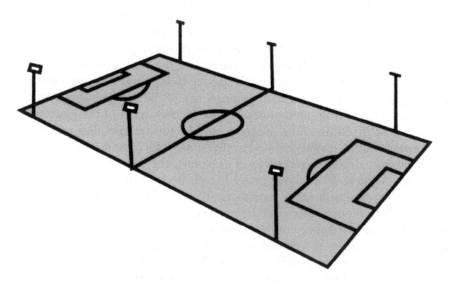

Figure A.10 Typical floodlight layout for a football pitch.

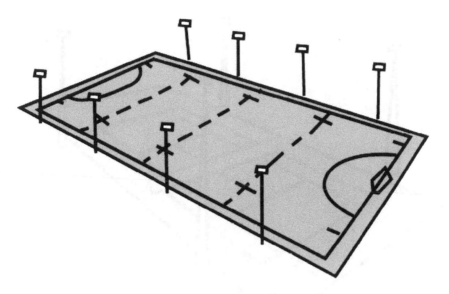

Figure A.11 Typical floodlight layout for a hockey pitch.

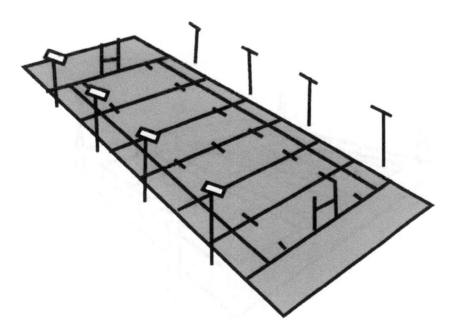

Figure A.12 Typical floodlight layout for an American football pitch.

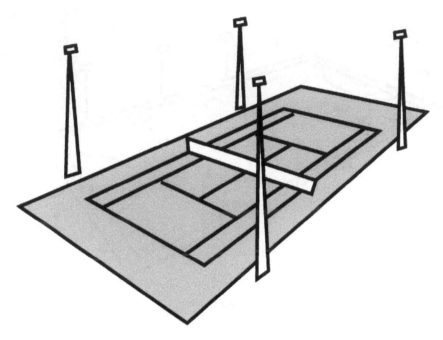

Figure A.13 Typical floodlight layout for a single tennis court.

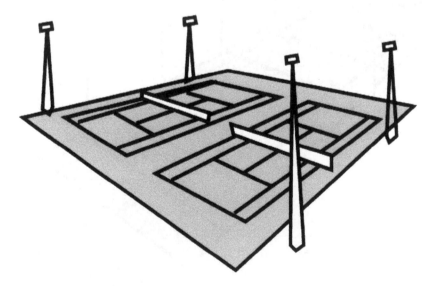

Figure A.14 Typical floodlight layout for a double tennis court.

A.3.3   Electrical considerations for festival lighting

A.3.3.1   Introduction

All electrical work must be carried out in accordance with the requirements of BS 7671 and be electrically inspected and tested before commissioning and at regular intervals thereafter to ensure the safety of operatives, the public and the installation.

A.3.3.2   Power supplies

Permanent dedicated power supplies are the preferred method of electrical supply for safety and convenience. Power supplies should only be taken from supply positions agreed with the client.

All controls and isolators should be accessible and local to the plugs and sockets or equipment they control. Plugs and sockets should comply with BS EN 60309, have a minimum IP rating of 67 and be installed at decoration height for safety and access.

Holes should not be drilled in lighting columns for a supply cable before the structural condition of the lighting column has been ascertained, and shall be no greater than 20 mm in diameter. Holes should be tapped and fitted with a cable compression gland. The cable compression gland shall be sealed or removed and replaced with a blanking plug when the supply cable is removed. Each lighting column should be treated on an individual basis and the size, location and details of the hole confirmed with the lighting column manufacturer to ensure structural adequacy.

Any private generating equipment should be 'silent running' and be electrically and structurally protected from damage and faults and unauthorised access. All fuel should be stored in purpose-made storage vessels which shall be protected from unauthorised access and damage. Suitable fire extinguishers shall be available at all times as shall suitable equipment to prevent any fuel spillages from contaminating the environment.

A.3.3.3   Earthing arrangements

Where electrical supplies are privately generated or where the supply authority does not provide an earth terminal, earth electrodes shall be installed in accordance with BS 7430 to ensure a disconnection time not greater than 0.2 s. Earth bonding must be provided to all structures such as stages, marquees, stalls and scaffolding when electrical supplies are provided. Catenary cables must also be bonded to earth.

A.3.3.4   Protection against electric shock

Electrical supplies for festival lighting or other temporary structures must be protected by a 30-mA RCD at the supply point. The circuits must also be suitably fused as required by BS 7671. All RCD units shall be tested in accordance with and at the frequencies recommended in BS 7671.

The use of extra low voltage equipment is recommended where equipment is within reach of the public. Where equipment is mounted at ground level, for example Christmas trees and tableaux, suitable rigid barriers should also be provided, to prevent access by the public.

A.3.3.5 Supply Authority requirements

The electricity company must be advised of all festival decorations being installed on unmetered supplies. Full details relating to the electrical load, burning hours and duration of the installation must be provided to enable the consumption of energy to be calculated and agreed.

A.3.3.6 Clearances

Festival lighting and decorations should be installed to provide a minimum clearance of 5.8 m over areas accessible to vehicles and 2.5 m elsewhere. These clearances must take into account the weight and wind loading of the decoration, any expansions of supporting cables and/or catenaries and designed for bending of support columns and posts.

A.3.4 Structural considerations for the erection of festival lighting

Festival lighting shall only be attached to lighting columns where the lighting columns have been tested or calculated as being structurally strong enough to support the weight and windage of the decoration.

Lighting columns, brackets, the foundations of planted root lighting columns, anchorages and attachments system for flange plates shall be designed to take account of the additional loadings imposed by the festival lighting. This is especially important for catenary decorations or power supply cables when the weight, span and sag will have to be carefully controlled.

The client for the festival lighting must ensure that all decorations are correctly supported and electrically safe and do not present a hazard to the public. To ensure the safety of the festival lighting, the client must obtain detailed specifications, that is electrical loading, weights and wind loading of the equipment being proposed and ascertained that these details are within the overall specification for the installation.

In some cases a local authority may have made allowance for a festival lighting installation as part of their lighting system. Lighting columns in such installations may have been designed and installed with additional limiting factors relating to the fitting of festival lighting equipment, for example: 'The lighting column may be fitted with festival decorations with a maximum windage of "X" square metres, a solidity ratio of 60 per cent located "Y" metres above ground level to the bottom of the decoration with a maximum offset of "Z" metres.' In addition a maximum electrical load may be specified. The client for the festival lighting shall provide full details of all decorations proving that they are within the design limitations specified.

Where the client installs their own supporting poles or columns on the public highway, they shall submit to the Highway Authority a copy of the Manufacturer's Design Certificate and, if required, Consultant's Check Certificate confirming compliance with the specification and design limitations for each type of lighting column or pole proposed. This certification should take account of all attachments including brackets, luminaires festival decorations, hanging baskets and so on.

Existing lighting columns should be routinely checked and certified in accordance with the recommendations in the ILE Technical Report No. 22 to ensure that they still comply with the structural requirements.

A.3.5 Structural considerations for the erection of catenary cables (for festival lighting)

Wall fixings and anchors shall be load tested and appropriate certificates issued each year prior to the fitting of any catenaries, cables or festival decorations. Such checks should include the integrity of the building or structure into which the fixings are made. A minimum load test for such fixings would be 5 kN; however, the advice of a structural engineer should be sought for each individual situation to ensure the correct test load has been applied.

Catenary systems should be designed and installed for the specific festival lighting or decoration to be supported and should be stainless steel or galvanised wire with a minimum cross-sectional area of at least 4 mm^2. Decorations and supply cables must be securely fixed to the catenary wire. Catenary power cables are usually suspended from wire catenary cables, but some cables are designed to be self-supporting up to a limiting span and minimum sag. Due to the higher risks involved in catenary cables spanning roads or pedestrian areas, the importance of regular inspection and testing cannot be overstated, whether the installation is permanent or erected for an annual festival (see also Section 3.6.11).

Where decorations and/or supply cables are suspended between lighting columns, there will be significant loading on the lighting columns dependent upon the weight, span and sag of the catenary in the direction of pull of the support wires and an increase in the wind loadings at right angles to the wires dependent upon windage area and solidity. These loads will result in bending of the lighting column even on lighting columns specifically designed for the purpose. Bending may be reduced by using stays, or further unloaded catenaries from a building. However, stays which provide horizontal resistance to the catenary load will also result in significant vertical buckling loads on the lighting column and should have as shallow an angle to the horizontal as possible.

There will be different structural considerations for different span layouts in plan. The loading from two catenaries on a single lighting column at right angles to each other is very different from two directly opposed spans, that is at 180° to each other. Equally the loading from three catenaries at 120° will be less. It is, therefore, essential that lighting columns are specifically designed for these loadings or existing lighting columns are checked and certified as being structurally satisfactory by the manufacturer or designer.

A.3.6　Power supplies for surface car parks

Electrical power to surface car park lighting is generally provided by dedicated underground private cable networks. The use of a loop in/out underground cable network where the power cables are looped in and out of each lighting column is preferable to a power network where service cables feeding individual lighting columns are jointed on to a main distribution cable. This is because it allows easier tracing of cable faults and maintenance on the network. The looping in/out of three-phase cables is not recommended due to the possibility of exposure to dangerous voltages under fault or damage conditions.

The use of a continuous underground duct system for the cable network is recommended as this allows for easier and simpler replacement of damaged or faulty cables without the need for extensive excavations which may disrupt the operation and use of the vehicle park.

All electrical distribution systems should be designed, installed and maintained in accordance with BS 7671.

Glossary

Absorptance (α)
Ratio of the absorbed radiant or luminous flux to the incident flux under specified conditions. (*See also CIE 17.4:1987; 845-04-75.*)

Access zone (of a tunnel) (*as per BS 5489-2:2003*)
Part of the open road immediately outside (in front of) an entrance portal, covering the distance over which an approaching driver should be able to see into the tunnel.

Access zone length (of a tunnel) (*as per BS 5489-2:2003*)
Distance between the stopping distance point ahead of an entrance portal and the entrance portal itself.

Adaptation (*as per BS EN 12665:2002*)
Process by which the state of the visual system is modified by previous and present exposure to stimuli that may have various luminances, spectral distributions and angular subtenses. (*See also CIE 7.4; 845-02-07.*)

Notes

1   The terms 'light adaptation' and 'dark adaptation' are also used, the former when the luminances of the stimuli are of at least several candelas per square metre, and the latter when the luminances are of less than some hundredths of a candela per square metre.
2   Adaptation to specific spatial frequencies, orientations, sizes, etc. are recognised as being included in this definition.

Ad-hoc lamp renewal
A maintenance procedure where individual lamps are renewed only when they fail. Particularly with discharge lamps, this is likely to result in a large proportion of the lamps operating well below their optimum economic efficacy and therefore a greater number of luminaires will need to be installed

to achieve the required maintained illuminance during the operational life of the installation. Some ad-hoc lamp renewal may be necessary when group lamp renewal is adopted to replace early lamp failures occurring between initial installation and the group lamp renewal.

Arc discharge; electric arc (*as per CIE 17.4:1987; 845-07-16*)
Electric discharge characterised by a cathode fall which is small compared with that in a glow discharge.

Note: The emission of the cathode results from various causes (thermionic emission, field emission, etc.) acting simultaneously or separately, but secondary emission plays only a small part.

Arc lamp (*as per CIE 17.4:1987; 845-07-33*)
A discharge lamp in which light is emitted by an arc discharge and/or by its electrodes.

Note: The electrodes may be either of carbon (operating in air) or of metal.

Arc tube (*as per CIE 17.4:1987; 845-08-28*)
The enclosure in which the arc of the lamp is confined.

Arrangement
The pattern in which the luminaires are sited in plan (e.g. regular, irregular, square, along one edge and around the perimeter, or for roads, staggered, opposite, single side, twin central, catenary).

Artificial sky glow
That part of the sky glow which is attributable to man-made sources of luminous radiation (e.g. outdoor electric lighting), including radiation that is emitted directly upwards and radiation that is reflected from the surface of the earth.

Average illuminance ($\bar{E}$) (*as per BS EN 12665:2002*)
Illuminance averaged over the specified area.

Unit: $lx = lm/m^2$.

Note: In practice this may be derived either from the total luminous flux falling on the surface divided by the total area of the surface, or alternatively from an average of the illuminances at a representative number of points on the surface.

Average illuminance (on a road area) ($\bar{E}_{hs}$) (*as per EN 13201-2:2003*)
Horizontal illuminance averaged over a road area.

Unit: lux (lx).

Average luminance ($\bar{L}$)
Luminance averaged over the specified area or solid angle.

Unit: $cd\,m^{-2}$.

Average road surface luminance (of a carriageway) ($\bar{L}$) (*as per EN 13201-2:2003*)
Luminance of the road surface averaged over the carriageway.

Note: Unit is candelas per square metre ($cd\,m^{-2}$).

Baffle
A single element, normally fitted inside a luminaire, in order to mask or shield undesirable views of the lamp or its reflected image.

Ballast (*as per BS EN 12665:2002*)
Device connected between the supply and one or more discharge lamps which serves mainly to limit the current of the lamp(s) to the required value. (*See also CIE 17.4:1987; 845-08-34.*)

Note: A ballast may also include means of transforming the supply voltage, correcting the power factor and, either alone or in combination with a starting device, provide the necessary conditions for starting the lamp(s).

Ballast lumen factor (*as per BS EN 12665:2002*)
Ratio of the luminous flux emitted by a reference lamp when operated with a particular production ballast to the luminous flux emitted by the same lamp when operated with its reference ballast. (*See also CIE 17.4:1987; 845-09-63.*)

Base plate (of a lighting column)
Plate below ground level fixed to a planted lighting column to prevent the lighting column sinking into the ground and to help prevent the lighting column overturning.

Beacon globe
Amber globe fixed on or bracketed off a Belisha beacon post.

Beam angle
The total angle over which the luminous intensity of a beam drops to 50 per cent of the peak value.

Beam toe (for a road luminaire)
The angle in plan between the centre(s) of a (the) beam(s) and the road axis.

Belisha beacon post
Black and white striped post carrying a beacon globe and used to define a pedestrian crossing (Belisha crossing; Zebra crossing).

Blended lamp; self-ballasted mercury lamp (USA) (as per CIE 17.4:1987; 845-07-21)
Lamp containing in the same bulb a mercury vapour lamp and an incandescent lamp filament connected in series.

Note: The bulb may be diffusing or coated with phosphors.

Bowl (as per CIE 17.4:1987; 845-10-34)
Diffuser, refractor or transmitter in the form of a bowl, intended to be placed below the lamp.

Bracket
A component used to support a luminaire at a definite distance from the axis of the lower straight position of a column, of single, double or multiple form and integral with, or demountable from, the column.

Bracket projection
The horizontal distance for column or wall-mounted luminaires between the centreline of the column at ground level, or wall bracket mounting surface and the entry-point of the luminaire.

Bracket fixing
The connecting part on a column for securing a separate bracket. It may be of the same size or a different cross section from the lighting column.

Brightness contrast (as per BS EN 12665:2002)
Subjective assessment of the difference in brightness between two or more surfaces seen simultaneously or successively.

Brightness: Luminosity (obsolete) (as per BS EN 12665:2002)
Attribute of a visual sensation according to which an area appears to emit more or less light. (See also CIE 17.4:1987; 845-02-28.)

Bulb
Transparent or transluscent gas-tight envelope enclosing the luminous element(s) of a lamp. Bulb – (deprecated) a term commonly used instead of lamp.

Bulkhead luminaire (as per CIE 17.4:1987; 845-10-16)
Protected luminaire of compact design intended to be fixed directly on a vertical or horizontal surface.

Bulk lamp renewal
See Group lamp renewal.

Cable entry slot (of a lighting column)
Opening in the lighting column below the ground for the cable entry.

Calculation grid
A series of calculation points set out in a grid formation.

Calculation point
A point on the working plane where a photometric calculation is to be made.

Candela (cd) (*as per CIE 17.4:1987; 845-01-50*)
SI unit of luminous intensity: The candela is the luminous intensity, in a given direction, of a source that emits monochromatic radiation of frequency 540×10^{12} Hz and that has a radiant intensity in that direction of 1/683 W/sr.

Unit: cd = lm/sr.

Carriageway
That part of the road normally used by vehicular traffic.

Chroma
In the Munsell system, an index of saturation of colour, ranging from 0 for neutral grey to 10 or over for strong colours. A low chroma implies a pastel shade.

Chromatic (perceived) colour (*as per CIE 17.4:1987; 845-02-27*)
In the perceptual sense: Perceived colour possessing hue. In everyday speech, the word 'colour' is often used in this sense in contradistinction to white, grey or black. The adjective 'coloured' usually refers to chromatic colour.

In the psychophysical sense: *See chromatic stimulus.*

Chromatic stimulus (*as per CIE 17.4:1987; 845-03-07*)
A stimulus that, under the prevailing conditions of adaptation, gives rise to a chromatic perceived colour.

Note: In the colorimetry of object colour, stimuli having purities greater than zero are usually considered to be chromatic stimuli.

Chromaticity *(as per BS EN 12665:2002)*
Property of a colour stimulus defined by its chromaticity coordinates, or by its dominant or complimentary wavelength and purity taken together. *(See also CIE 17.4:1987; 845-03-34.)*

Chromaticity co-ordinates *(as per BS EN 12665:2002)*
Ratio of each of a set of three tristimulus values to their sum. *(See also CIE 17.4:1987; 845-03-33.)*

Notes

1   As the sum of the three chromaticity co-ordinates equals 1, two of them are sufficient to define a chromaticity.
2   In the CIE standard colorimetric systems, the chromaticity co-ordinates are presented by the symbols x, y, z and x_{10}, y_{10}, z_{10}.

CIE standard photometric observer *(as per CIE 17.4:1987; 845-01-23)*
Ideal observer having a relative spectral responsivity curve that conforms to the $V(\lambda)$ function for photopic vision or to the $V'(\lambda)$ function for scotopic vision, and that complies with the summation law implied in the definition of luminous flux.

Cold cathode lamp *(as per CIE 17.4:1987; 845-07-27)*
A discharge lamp in which light is produced by the positive column of a glow discharge.

Note: Such a lamp is generally fed from a device providing sufficient voltage to initiate starting without special means.

Colorimeter *(as per BS EN 12665:2002)*
Instrument for measuring colorimetric quantities, such as the tristimulus values of a colour stimulus.

Colorimetry *(as per CIE 17.4:1987; 845-05-10)*
Measurement of colours based on a set of conventions.

Colour appearance
The apparent colour of light emitted by a particular light source – often expressed in terms of 'cool' (4000 K), 'intermediate' (3500 K) and 'warm' (3000 K).

Note: The terms 'cool', 'intermediate' and 'warm', together with the values expressed above, are only generalisations but are commonly used for descriptions of colour appearance.

Colour contrast (*as per BS EN 12665:2002*)
Subjective assessment of the difference in colour between two or more surfaces seen simultaneously or successively.

Colour rendering (*as per BS EN 12665:2002*)
Effect of an illuminant on the colour appearance of objects by conscious or subconscious comparison with their colour appearance under a reference illuminant. (*See also CIE 17.4:1987; 845-02-59.*)

Colour rendering index (*R_a*) (*as per CIE 17.4:1987; 845-02-61*)
Measure of the degree to which the psychophysical colours of an object illuminated by the test illuminant conform to that of the same object illuminated by the reference illuminant, suitable allowance having been made for the state of chromatic adaptation.

Colour stimulus (*as per BS EN 12665:2002*)
Visible radiation entering the eye and producing a sensation of colour, either chromatically or achromatically. (*See also CIE 17.4:1987; 845-03-02.*)

Colour temperature (*T_c*)
Temperature of a Planckian radiator whose radiation has the same chromaticity as that of a given stimulus. (*See also CIE 17.4:1987; 845-03-49.*)

Unit: kelvin (K).

Note: The reciprocal colour temperature is also used, unit is K^{-1}.

Column
See Lighting column.

Conflict area
Areas such as shopping streets, road intersections of some complexity, roundabouts, queuing areas, pedestrian crossings, etc., where different classes of road users or different traffic flows meet and/or cross.

Contrast (*as per BS EN 12665:2002*)
In the perceptual sense: Assessment of the difference in appearance of two or more parts of a field seen simultaneously or successively (hence: brightness contrast, lightness contrast, colour contrast, simultaneous contrast, successive contrast, etc.).

In the physical sense: Quantity intended to correlate with the perceived brightness contrast, usually defined by one of a number of formulae which involve the luminances of the stimuli considered, e.g.: $\Delta L/L$ near the luminance threshold, or L_1/L_2 for much higher luminances. (*See also CIE 17.4:1987; 845-02-47.*)

Correlated colour temperature (T_{cp}) (*as per BS EN 12665:2002*)
Temperature of the Planckian radiator whose perceived colour most closely resembles that of a given stimulus at the same brightness and under specified viewing conditions. (*See also CIE 17.4:1987; 845-03-50.*)

Unit: kelvin (K).

Notes

1 The recommended method of calculating the CCT of a stimulus is to determine on a chromaticity diagram the temperature corresponding to the point on the Planckian locus that is intersected by the agreed isotemperature line containing the point representing the stimulus.
2 Reciprocal CCT is used rather than reciprocal colour temperature whenever CCT is appropriate.

Cosine correction (*as per BS EN 12665:2002*)
Correction of a detector for the influence of the incident direction of the light.

Note: For the ideal detector, the measured illuminance is proportional to the cosine of the angle of incidence of the light. The angle of incidence is the angle between the direction of the light and the normal to the surface of the detector.

Cowl
A form of shield fitted to the outside of a luminaire to minimise obtrusive light, normally in the form of a hood to limit upward light.

Critical flicker frequency
See Fusion frequency.

Curfew
The time after which stricter requirements (e.g. for the control of obtrusive light) will apply; often a condition of use of lighting applied by a planning authority.

Cut-off (*as per CIE 17.4:1987; 845-10-29*)
Technique used for concealing lamps and surfaces of high luminance from direct view in order to reduce glare.

Note: In public lighting, distinction is made between full-cut-off luminaires, semi-cut-off luminaires and non-cut-off luminaires.

Cut-off angle (of a luminaire) (*as per CIE 17.4:1987; 845-10-29*)
Angle, measured up from nadir, between the vertical axis and the first line of sight at which the lamps and the surfaces of high luminance are not visible. For a floodlight this angle is usually measured from the beam axis.

Cut off luminaire (road lighting)
A classification applied to a luminaire which, when installed in a specified design attitude the maximum permissible value of intensity emitted does not exceed 10 cd/1000 lm at 90° (up to a maximum of 1000 cd whatever the luminous flux emitted) and 30 cd/1000 lm at 80°.

Cycle lane
A part of a carriageway allocated for use by cyclists.

Cycle route
A route for cyclists which may include cycle tracks, cycle lanes and other public highways.

Cycle track
A way comprised in or constituting a highway with a right of way for pedal cycles with, or without, a right of way on foot.

Cylindrical illuminance (at a point, for a direction) (*E_z*) (*as per BS EN 12665:2002*)
Total luminous flux falling on the curved surface of a very small cylinder located at the specified point divided by the curved surface area of the cylinder. The axis of the cylinder is taken to be vertical unless stated otherwise. (*See also CIE 17.4:1987; 845-01-41.*)

Unit: lux = lm/m^2.

Quantity defined by the formula: $E_z = 1/\pi \int_{4\pi sr} L \sin \varepsilon \, d\Omega$

where $d\Omega$ is the solid angle of each elementary beam passing through the given point; L its luminance at that point; and ε the angle between it and the given direction.

Dark adaptation
The state of the visual system when it has become adapted to a very low luminance (less than some hundredths of a candela per square metre) usually associated with scotopic vision.

Daylight (*as per BS EN 12665:2002*)
Visible part of global solar radiation.

Note: When dealing with actinic effects of optical radiations, this term is commonly used for radiations extending beyond the visible region of the spectrum.

Daylight factor (*D*) (*as per BS EN 12665:2002*)
Ratio of the illuminance at a point on a given plane due to the light received directly or indirectly from a sky of assumed or known luminance distribution, to the illuminance on a horizontal plane due to an unobstructed hemisphere of this sky. The contribution of direct sunlight to both illuminances is excluded. (*See also CIE 17.4:1987; 845-09-97.*)

Notes

1   Glazing, dirt effects, etc. are included.
2   When calculating the lighting of interiors, the contribution of direct sunlight must be considered separately.

Daylight screen (of a tunnel) (*as per BS 5489-2:2003*)
Devices that transmit (part of) the ambient daylight.

Note: Daylight screens may be applied for the lighting of the threshold zone of a tunnel.

Defined area
The area over which the lighting calculations are to be made.

Design speed (*as per BS 5489-2:2003*)
Speed adopted for a particular stated purpose in designing a road.

Diffuse reflection (*as per CIE 17.4:1987; 845-04-47*)
Diffusion by reflection in which, on a macroscopic scale, there is no regular reflection.

Diffuser (*as per CIE 17.4:1987; 845-04-53*)
Device used to alter the spatial distribution of radiation and depending essentially on the phenomenon of diffusion.

Note: If all the radiation reflected or transmitted by the diffuser is diffused with no regular reflection or transmission, the diffuser is said to be completely diffusing, independently of whether or not the reflection or transmission is isotropic.

Diffused lighting (*as per BS EN 12665:2002*)
Lighting in which the light on the working plane or on an object is not incident predominantly from a particular direction. (*See also CIE 17.4:1987; 845-09-20.*)

Diffusion; scattering (*as per CIE 17.4:1987; 845-04-43*)
Process by which the spatial distribution of a beam of radiation is changed when it is deviated in many directions by a surface or by a medium, without change of frequency of its monochromatic components.

Note: A distinction is made between selective diffusion and non-selective diffusion according to whether or not the diffusion properties vary with the wavelength of the incident radiation.

Direct lighting (*as per BS EN 12665:2002*)
Lighting by means of luminaires having a distribution of luminous intensity such that the fraction of the emitted luminous flux directly reaching the working plane, assumed to be unbounded, is 90–100 per cent. (*See also CIE 17.4:1987; 845-09-14.*)

Direct solar radiation (*as per BS EN 12665:2002*)
That part of the extraterrestrial solar radiation which as a collimated beam reaches the Earth's surface after selective attenuation by the atmosphere. (*See also CIE 17.4:1987; 845-09-79.*)

Directional lighting (*as per BS EN 12665:2002*)
Lighting in which the light on the working plane or on an object is incident predominantly from a particular direction. (*See also CIE 17.4:1987; 845-09-19.*)

Disability glare (*as per BS EN 12665:2002*)
Glare that impairs the vision of objects without necessarily causing discomfort. (*See also CIE 17.4:1987; 845-02-57.*)

Discomfort glare (*as per BS EN 12665:2002*)
Glare that causes discomfort without necessarily impairing the vision of objects. (*See also CIE 17.4:1987; 845-02-56.*)

Discharge lamp (*as per CIE 17.4:1987; 845-07-17*)
Lamp in which the light is produced, directly or indirectly, by an electrical discharge through a gas, a metal vapour or a mixture of several gases and vapours.

Note: According by as the light is mainly produced in a gas or in a metal vapour, one distinguishes between gaseous discharge lamps, e.g. xenon, neon, helium, nitrogen, carbon dioxide lamp, and metal vapour lamps, such as the mercury vapour lamp and the sodium vapour lamp.

Door opening (of a lighting column)
Opening in the lighting column for access to electrical equipment.

Downward light output ratio (of a luminaire) (*as per CIE 17.4:1987; 845-09-40*)
Ratio of the downward flux of the luminaire, measured under specified practical conditions with its own lamps and equipment, to the sum of the individual luminous fluxes of the same lamps when operated outside the luminaire with the same equipment, under specified conditions.

Note: For luminaires using incandescent lamps only, the optical light output ratio and the light output ratio are the same in practice.

Efficacy
See Luminous efficacy.

Electric discharge (in a gas) (*as per CIE 17.4:1987; 845-07-11*)
The passage of an electric current through gases and vapours by the production and movements of charge carriers under the influence of an electric field.

Note: The phenomenon results in the emission of electromagnetic radiation which plays an essential part in all its applications in lighting.

Electricity company
The company responsible for the distribution of electrical energy in a predefined area.

Emergency lane (hard shoulder)
A lane parallel to the traffic lane(s), not intended for normal traffic, but for emergency (police) vehicles and/or broken down vehicles.

Emergency lighting (*as per BS EN 12665:2002*)
Lighting provided for use when the supply to the normal lighting fails. (*See also CIE 17.4:1987; 845-09-10.*)

Energy efficiency
See Installed efficacy and Installed power density.

Entrance portal (of a tunnel) (*as per BS 5489-2:2003*)
Part of the tunnel construction that corresponds to the beginning of the covered part of the tunnel or, when open daylight screens are used, to the beginning of the daylight screens.

Entrance zone (of a tunnel) (*as per BS 5489-2:2003*)
Combination of the threshold zone and the transition zone(s).

Environmental zone
Designated zone (area) where additional controls on exterior lighting are imposed by local or national authorities.

Escape lighting (*as per CIE 17.4:1987; 845-09-11*)
That part of Emergency lighting provided to ensure that an escape route can be effectively identified and used.

Exit portal (of a tunnel) (*as per BS 5489-2:2003*)
End of the covered part of the tunnel or, when daylight screens are used, end of the daylight screens.

Exit zone (of a tunnel) (*as per BS 5489-2:2003*)
Part of the tunnel where, during daytime, the vision of the driver approaching the exit is influenced predominantly by the brightness outside the tunnel.

Note: The exit zone stretches from the end of the interior zone to the exit portal of the tunnel.

Exitance
See Radiant exitance.

Explosion
An uncontrolled combustion wave.

Explosive atmosphere
An atmosphere where flammable materials in the form of gases, vapours, mists or dusts are mixed with air.

Fixed sign (*as per BS EN 12899-1:2001*)
A sign which is intended to remain fixed in position and whose supports are usually set into the ground.

Flange plate (of a lighting column)
A plate, with an opening for cable entry, attached rigidly to a lighting column which is surface-mounted, to allow it to be secured to a concrete foundation or to other structures.

Flashing light (*as per CIE 17.4:1987; 845-11-11*)
Rhythmic light in which every appearance of the light (flash) is of the same duration, and, except possibly for rhythms with rapid rates of flashing, the total duration of light in a period is clearly shorter than the total duration of darkness.

Note: The term eclipse is used for the interval of darkness between two successive appearances of light.

Flashed (luminous) area
See Luminous area.

Flicker (*as per BS EN 12665:2002*)
Impression of unsteadiness of visual sensation induced by a light stimulus whose luminance or spectral distribution fluctuates with time. (*See also CIE 17.4:1987; 845-09-04.*)

Floodlight (*as per CIE 17.4:1987; 845-10-28*)
Projector designed for floodlighting, usually capable of being pointed in any direction.

Floodlighting (*as per BS EN 12665:2002*)
Lighting of a scene or object, usually by projectors, in order to increase considerably its illuminance relative to its surroundings.

Fluorescent lamp (*as per CIE 17.4:1987; 845-07-26*)
A discharge lamp of the low pressure mercury type in which most of the light is emitted by one or several layers of phosphors excited by the UV radiation from the discharge.

Note: These lamps are frequently tubular and, in the UK, are then usually called fluorescent tubes.

Footpath (*as per BS 5489: Part 3; 1992*)
A means of passage for pedestrians. Footpaths may be across open spaces, or between buildings.

Footway (*as per BS 5489: Part 3; 1992*)
The portion of a road that is reserved for pedestrians.

Frog (for road lighting luminaires)
A mounting device for attaching a post-top luminaire (usually of heritage-style and derived from gas light designs) to a post-top lighting column. The frog comprises of three or more slender arms attached to the luminaire base; the other end of the arms terminate in a socket which fits over the post-top fixing on the top of the lighting column shaft.

Full cut-off luminaire (road lighting)
A classification applied to a luminaire which, when installed in a specified design attitude, gives zero intensity at and above the horizontal.

Full radiator
A thermal radiator obeying Planck's radiation law and having the maximum possible radiant exitance for all wavelengths for a given temperature; also called a 'black body' to emphasise its absorption of all incident radiation.

Fusion frequency; critical flicker frequency (for a given set of conditions) (*as per BS EN 12665:2002*)
Frequency of alteration of stimuli above which flicker is not perceptible. (*See also CIE 17.4:1987; 845-02-50.*)

Gas filled (incandescent) lamp (*as per CIE 17.4:1987; 845-07-09*)
Incandescent lamp in which the luminous element operates in a bulb filled with an inert gas.

General colour rendering index (of a light source) (R_a)
Value intended to specify the degree to which objects illuminated by a light source have an expected colour relative to their colour under a reference light source. (*See also CIE 17.4:1987; 845-02-63.*)

Note: R_a is derived from the colour rendering indices for a specified set of eight test colour samples. R_a has a maximum of 100, which generally occurs when the spectral distributions of the light source and the reference light source are substantially identical. (*See also CIE Publication 13.2.*)

General diffused lighting (*as per BS EN 12665:2002*)
Lighting by means of luminaires having a distribution of luminous intensity such that the fraction of the emitted luminous flux directly reaching the working plane, assumed to be unbounded, is 40–60 per cent. (*See also CIE 17.4:1987; 845-09-16.*)

General lighting (*as per BS EN 12665:2002*)
Substantially uniform lighting of an area without provision for special local requirements. (*See also CIE 17.4:1987; 845-09-06.*)

Geometry (of a road lighting system) (*as per BS 5489-1:2003*)
Interrelated linear dimensions and characteristics of the road lighting system, i.e. spacing, mounting height, transverse position and arrangement.

Glare (*as per BS EN 12665:2002*)
Condition of vision in which there is discomfort or a reduction in the ability to see details or objects, caused by an unsuitable distribution or range of luminance, or to extreme contrasts. (*See also CIE 17.4:1987; 845-02-52.*) *See also Disability glare* and *Discomfort glare*.

Glare index system
A system which produces a numerical index calculated according to the method described in CIBSE *TM10*. It enables the discomfort glare from lighting installations to be ranked in the order of severity and the permissible limit of discomfort glare from an installation to be prescribed quantitatively.

Global solar radiation (*as per BS EN 12665:2002*)
Combined direct solar radiation and diffuse sky radiation.

Globe (*as per CIE 17.4:1987; 845-10-36*)
Envelope of transparent or diffusing material, intended to protect the lamp, to diffuse the light, or to change the colour of the light.

Glow discharge (*as per CIE 17.4:1987; 845-07-12*)
Electrical discharge in which the secondary emission from the cathode is much greater than the thermionic emission.

Note: This discharge is characterised by a considerable cathode fall (typically 70 V or more) and by low current density at the cathode (some $10 A/m^2$).

Goniophotometer (*as per CIE 17.4:1987; 845-05-22*)
Photometer for measuring the directional light-distribution characteristics of sources, luminaires, media or surfaces.

Group lamp renewal
A maintenance procedure where all lamps in an installation are renewed at one time. The lumen maintenance characteristics and probability of lamp failure dictate the period after which group renewal, usually linked with luminaire cleaning, will take place. This method has visual, electrical and financial advantages over the alternative of 'ad-hoc renewal'.

Hard shoulder
See Emergency lane.

Hazardous area
An area where a flammable substance in the form of a gas vapour or dust mixed with air is present and in such a concentration that it can explode if it comes into contact with an ignition source.

Hazardous environment
An environment in which there exists risk of fire or explosion.

Hemispherical illuminance (at a point) (E_{hs}) (*as per BS EN 12665:2002*)
Total luminous flux failing on the curved surface of a very small hemisphere located at the specified point divided by the curved surface area of the hemisphere. The base of the hemisphere is taken to be horizontal unless stated otherwise.

Unit: lux $= lm/m^2$.

High intensity discharge lamp; HID lamp (*as per CIE 17.4:1987; 845-07-19*)
An electric discharge lamp in which the light-producing arc is stabilised by wall temperature and the arc has a bulb wall loading in excess of $3 W/cm^2$.

Note: HID lamps include groups of lamps known as high pressure mercury, metal halide and high pressure sodium lamps.

High mast
A lighting column of height 18 m or greater supporting luminaires.

High pressure mercury (vapour) lamp (*as per CIE 17.4:1987; 845-07-20*)
A high intensity discharge lamp in which the major portion of the light is produced, directly or indirectly, by radiation from mercury operating at a partial pressure in excess of $100 kP_a$.

Note: This term covers clear, phosphor-coated (mercury fluorescent) and blended lamps. In a fluorescent mercury-discharge lamp, the light is produced partly by the mercury vapour and partly by a layer of phosphors excited by the UV radiation of the discharge.

High pressure sodium (vapour) lamp (*as per CIE 17.4:1987; 845-07-23*)
A high intensity discharge lamp in which the major portion of the light is produced, directly or indirectly, by radiation from sodium operating at a partial pressure in excess of $10 kP_a$.

Note: The term covers lamps with clear or diffusing bulbs.

Homezone
A geographical area with a high percentage of residential to other usage of buildings and a density exceeding 20 residence units per hectare.

Horizontal illuminance
The illuminance on a horizontal plane at a specified height (ground level, unless otherwise specified).

Hostile environment
An environment in which the lighting equipment may be subject to chemical, thermal or mechanical attack.

Hot cathode lamp *(as per CIE 17.4:1987; 845-07-28)*
A discharge lamp in which the light is produced by the positive column of an arc discharge.

Note: Such a lamp generally requires a special starting device or circuit.

Hue *(as per CIE 17.4:1987; 845-02-35)*
Attribute of a visual sensation according to which an area appears to be similar to one of the perceived colour, red, yellow, green and blue, or to a combination of two of them.

***I* Table**
For a luminaire, the table of intensities (in candelas per 1000 lamp lumens) that describes the light distribution of the luminaire. Intensities are tabulated at defined angular intervals (horizontal and vertical) and are often associated with lines of text and numeric information describing other features of the luminaire in a specific format that can be read by appropriate computer programs to perform installation calculations.

Ignitor *(as per CIE 17.4:1987; 845-08-33)*
A device intended, either by itself or in combination with other components in the circuit, to generate voltage pulses to start a discharge lamp without providing preheating of the electrodes.

Illuminance (at a point of a surface) (E) *(as per BS EN 12665:2002)*
Quotient of the luminous flux $d\Phi$ incident on an element of the surface containing the point, by the area dA of that element. (*See also CIE 17.4:1987; 845-01-38.*)

Equivalent definition: Integral, taken over the hemisphere visible from the given point, of the expression

$$L \cos \Theta \, d\Omega$$

where L is the luminance at the given point in the various directions of the incident elementary beams of solid angle dΩ; and Θ is the angle between any of these beams and the normal to the surface at the given point.

$$E = \frac{d\Theta}{dA} = \int\limits_{2\pi sr} L \cos\Theta d\Omega$$

Unit: lx = lm/m^2.

See also *Average illuminance, Maintained illuminance, Maximum illuminance and Minimum illuminance.*

Illuminance uniformity (*as per BS EN 12665:2002*)
Ratio of minimum illuminance to average illuminance on a surface. (*See also CIE 17.4:1987; 845-09-58 uniformity ratio of illuminance.*)

Note: Use is also made of the ratio of minimum illuminance to maximum illuminance in which case this should be specified explicitly.

Illuminant (*as per CIE 17.4:1987; 845-03-10*)
Radiation with a relative spectral power distribution defined over the wavelength range that influences object colour perception.

Note: In everyday English, this term is not restricted to this sense, but is also used for any kind of light falling on a body or scene.

Illumination
See Lighting.

Incandescence (*as per CIE 17.4:1987; 845-04-15*)
Emission of optical radiation by process of thermal radiation.

Note: In the USA incandescence is restricted to visible radiation.

Incandescent (electric) lamp (*as per CIE 17.4:1987; 845-07-04*)
Lamp in which light is produced by means of an element heated to incandescence by the passage of an electric current.

Indirect lighting (*as per CIE 17.4:1987; 845-09-18*)
Lighting by means of luminaires having a distribution of luminous intensity such that the fraction of the emitted luminous flux directly reaching the workplane, assumed to be unbounded, is 0–10 per cent.

Ingress protection (IP) number

A two-digit number associated with a luminaire or other enclosure. The first digit classifies the degree of protection provided against the ingress of dust or solid objects. The second digit classifies the degree of protection provided against the ingress of moisture. A full specification of IP rating is given in BS EN 60529:1992, Specification for degrees of protection provided by enclosures (IP code).

Initial illuminance (E_i) (as per BS EN 12665:2002)

Average illuminance when the installation is new.

Unit: $\text{lx} = \text{lm/m}^2$.

Initial light output

The luminous flux from a new lamp. In the case of discharge lamps this is usually the output after 100 hours of operation.

Unit: lm.

Initial luminance (L_i) (as per BS EN 12665:2002)

Average luminance when the installation is new.

Installed efficacy

A factor which quantifies the effectiveness of a lighting installation in converting electrical power to light. Specifically, it is the product of the lamp circuit luminous efficacy and the utilisation factor.

Unit: lm/W.

Installed loading (as per BS EN 12665:2002)

Installed power of the lighting installation per unit area (for interior and exterior areas) or per unit length (for road lighting).

Unit: W/m^2 (for area) or kW/km (for road lighting).

Installed power density

The installed power density per 100 lux is the power needed per square metre of floor area to achieve 100 lux on a horizontal plane with general lighting.

Integrating sphere; Ulbricht sphere (as per CIE 17.4:1987; 845-05-24)

Hollow sphere whose internal surface is diffuse reflector, as non-selective as possible.

Note: An integrating sphere is used frequently with a radiometer or photometer.

Integrating photometer (*as per CIE 17.4:1987; 845-05-25*)
Photometer for measuring luminous flux, generally incorporating an integrating sphere.

Intensity
See Luminous intensity.

Interior zone (of a tunnel) (*as per BS 5489-2:2003*)
Part of the tunnel following directly after the transition zone.

Note: The interior zone stretches from the end of the transition zone to the
     beginning of the exit zone.

Internally illuminated post
Traffic sign post where part of the post is cut-out to allow light from an internal light source to illuminate a translucent diffuser fixed across the cut-out section(s).

Irradiance (at a point of a surface) (E_e, E) (*as per CIE 17.4:1987; 845-01-37*)
Quotient of the radiant flux $d\Phi_e$ incident on an element of the surface containing the point, by the area dA of that element.

Equivalent definition: Integral, taken over the hemisphere visible from the given point, of the expression

$$L_e \cos \Theta \, d\Omega$$

where L_e is the radiance at the given point in the various directions of the incident elementary beam of solid angle $d\Omega$; and Θ is the angle between any of the beams and the normal to the surface at the given point.

$$E_e = \frac{d\Phi e}{dA} = \int_{2\pi sr} L_e \cos \Theta \, d\Omega$$

Unit: W/m².

Iso-illuminance curve; iso-illuminance line (USA); iso-lux curve or line
(deprecated) (*as per CIE 17.4:1987; 845-09-57*)
Locus of points on a surface where the illuminance has the same value.

Iso-intensity curve; iso-intensity line (USA); iso-candela curve or line
(deprecated) (of a source) (*as per CIE 17.4:1987; 845-09-28*)
Curve traced on a sphere that has its centre at the centre of the light source, joining all the points corresponding to those directions in which the luminous intensity is the same, or a plane projection of that curve.

Iso-intensity diagram; iso-candela diagram (deprecated) (*as per CIE 17.4:1987; 845-09-29*)
Array of iso-intensity curves.

Iso-illuminance diagram; iso-lux diagram (deprecated)
Array of iso-illuminance curves.

Iso-luminance curve (*as per CIE 17.4:1987; 845-09-56*)
Locus of points on a surface at which the luminance is the same, for given positions of the observer and of the source or sources in relation to the surface.

K Factor
A factor used to multiply the basic average wind pressure used in the calculation of lighting column loadings to allow for the variation in average wind pressure over the country and the exposure class.

Lamp
Source made in order to produce optical radiation, usually visible. (*See also CIE 17.4:1987; 845-07-03.*)

Note: This term is also sometimes incorrectly used for certain types of luminaires.

Lamp lumen maintenance factor (LLMF) (*as per BS EN 12665:2002*)
Ratio of the luminous flux of a lamp at a given time in its life to the initial luminous flux.

Lamp survival factor (LSF) (*as per BS EN 12665:2002*)
Fraction of the total number of lamps which continue to operate at a given time under defined conditions and switching frequency.

Lantern
See Luminaire.

Large car park
A vehicle park with a capacity of greater than 100 vehicles for major shopping centres, major sports centres or multipurpose building complexes.

Life (of a lamp) (*as per CIE 17.4:1987; 845-07-61*)
The total time for which a lamp has been operated before it becomes useless, or is considered to be so according to specified criteria.

Note: Lamp life is usually expressed in hours.

Life of lighting installation *(as per BS EN 12665:2002)*
Period after which the installation cannot be restored to satisfy the required performance because of non-recoverable deterioration.

Light *(as per CIE 17.4:1987; 845-01-06)*
Perceived light *(See also CIE 17.4:1987; 845-02-17.)*

Visible radiation *(See also CIE 17.4:1987; 845-01-03.)*

Notes

1   The word *light* is sometimes used in sense 2 for optical radiation extending outside the visible range, but this usage is not recommended.
2   The terms 'light' in English and 'Licht' in German are also used, especially in visual signalling, for certain lighting devices and for light signals.

Light adaptation
The state of the visual system when it has become adapted to a luminance of at least several candelas per square metre.

Light distribution (Alternatively, 'See *luminous intensity distribution*')
The distribution of luminous intensity from an individual luminaire in various directions in space.

Light loss factor; maintenance factor (obsolete) *(as per CIE 17.4:1987; 845-09-59)*
Ratio of the average illuminance on the working plane after a certain period of use of a lighting installation to the average illuminance obtained under the same conditions for the installation considered conventionally as new.

Notes

1   The term *depreciation factor* has been formerly used to designate the reciprocal of the above ratio.
2   The light losses take into account dirt accumulation on luminaires and room surfaces (if applicable) and lamp depreciation.

Light output ratio (of a luminaire); luminaire efficiency (USA) *(as per CIE 17.4:1987; 845-09-39)*
Ratio of the total flux of the luminaire, measured under specified practical conditions with its own lamps and equipment, to the sum of the individual luminous fluxes of the same lamps when operated outside the luminaire with the same equipment, under specified conditions.

Note: For luminaires using incandescent lamps only, the optical light output ratio and the light output ratio are the same in practice.

Light pollution
The spillage of light into areas where it is not desired.

Light trespass
Light, normally measured in illuminance, that impacts onto a surface outside of the area designed to be lit by the installation concerned.

Lighting; illumination *(as per CIE 17.4:1987; 845-09-01)*
Application of light to a scene, objects or their surroundings so that they may be seen.

Note: This term is also used colloquially with the meaning 'lighting system' or 'lighting installation'.

Lighting chain; lighting string (USA) *(as per CIE 17.4:1987; 845-10-24)*
Set of lamps arranged along a cable and connected in series or parallel.

Lighting column
Support intended to hold one or more luminaires, consisting of one or more parts: a post, possibly an extension piece and, if necessary, a bracket. It does not include columns for catenary lighting.

Load factor
The ratio of the energy actually consumed by a lighting installation with controls over a specified period of time to the energy that would have been consumed had the lighting installation been operated without controls during the same period of time.

Local car park
A car park with a capacity less than 20 vehicles for local shops, schools, churches or communal parking for residential properties.

Local lighting *(as per CIE 17.4:1987; 845-09-07)*
Lighting for a specific visual task, additional to and controlled separately from the general lighting.

Localised lighting *(as per CIE 17.4:1987; 845-09-08)*
Lighting designed to illuminate an area with a higher illuminance at certain specified positions, for instance those at which work is carried out.

Long-arc lamp *(as per CIE 17.4:1987; 845-07-35)*
An arc lamp, generally of high pressure, in which the distance between the electrodes is large, the arc filling the discharge tube and therefore being stabilised.

Longitudinal uniformity ratio (of road surface luminance of a driving lane) *(as per EN 13201-2:2003)*
Ratio of the lowest to highest road surface luminance along in a line in the centre along a driving lane.

Longitudinal uniformity ratio (of road surface luminance of a carriageway) *(U_l)* *(as per EN 13201-2:2003)*
Lowest of the longitudinal uniformities of the driving lanes of the carriageway.

Louvre
A multiple arrangement of baffles, usually fitted to the outside of a luminaire in order to minimise obtrusive light. They can be horizontal, vertical or circular in design.

Low pressure mercury (vapour) lamp *(as per CIE 17.4:1987; 845-07-22)*
A discharge lamp of the mercury vapour type, with or without a coating of phosphors, in which during operation the partial pressure of the vapour does not exceed 100 Pa.

Low pressure sodium (vapour) lamp *(as per CIE 17.4:1987; 845-07-24)*
A discharge lamp in which light is produced by radiation from sodium vapour operating at a partial pressure 0.1–1.5 Pa.

Lumen (lm) *(as per CIE 17.4:1987; 845-01-51)*
SI unit of luminous flux: Luminous flux emitted in unit solid angle (steradian) by a uniform point source having a luminous intensity of 1 cd.

Equivalent definition: Luminous flux of a beam of monochromatic radiation whose frequency is 540×10^{12} Hz and whose radiant flux is $1/683$ W.

Luminaire *(as per BS EN 12665:2002)*
Apparatus which distributes, filters or transforms the light transmitted from one or more lamps and which includes, except the lamps themselves, all the parts necessary for fixing and protecting the lamps and, where necessary, circuit auxiliaries together with the means for connecting them to the electric supply. (*See also* CIE 17.4:1987; 845-10-01.)

Note: The term 'lighting fitting' is deprecated.

Luminaire array (road lighting)
A group of luminaires on an individual high mast or lighting column.

Luminaire fixing (road lighting)
The connecting part on the end of a post-top lighting column or of a bracket for securing a luminaire. It may be the end of the lighting column or the bracket itself or an additional part having the same or a different cross-section from the lighting column or bracket.

Luminaire fixing angle (road lighting)
Angle between the axis of the luminaire and the horizontal.

Luminaire guard
Device, generally in the form of a grid, used to shield the protective glass of the luminaire against mechanical shock.

Luminaire maintenance factor (LMF) (*as per BS EN 12665:2002*)
Ratio of the light output ratio of a luminaire at a given time to the initial light output ratio.

Luminance (in a given direction, at a given point of a real or imaginary surface) (L_V; L)
Luminous flux per unit solid angle transmitted by an elementary beam passing through the given point and propagating in the given direction, divided by the area of a section of that beam normal to the direction of the beam and containing the given point. (*See also CIE 17.4:1987; 845-01-35.*)
   It can also be defined as:

The luminous intensity of the light emitted or reflected in a given direction from an element of the surface, divided by the area of the element projected in the same direction.
The illuminance produced by the beam of light on a surface normal to its direction, divided by the solid angle of the source as seen from the illuminated surface.
It is the physical measurement of the stimulus which produces the sensation of brightness.

   Unit: $\mathrm{cd\,m^{-2}} = \mathrm{lm/(m^2/sr)}$.

   Technically defined: Quantity defined by the formula

$$L = \frac{d\Phi}{dA \times \cos\Theta \times d\Omega}$$

where $d\Phi$ is the luminous flux transmitted by an elementary beam passing through the given point and propagating in the solid angle $d\Omega$ containing the given direction; dA is the area of a section of that beam containing the given point; and Θ is the angle between the normal to that section and the direction of the beam.

The relationship between luminance and illuminance is given by the equation.

$$\text{Luminance} = \frac{\text{illuminance} \times \text{reflectance ractor}}{\pi}$$

This equation applies to a matt surface. For a non-matt surface, the reflectance is replaced by the luminance factor.

Luminance contrast (*as per BS EN 12665:2000*)
Photometric quantity intended to correlate with brightness contrast, usually defined by one of a number of formulae which involve the luminances of the stimuli considered. (*See also CIE 17.4:1987; 845-02-47.*) (*See also Contrast.*)

Note: Luminance contrast may be defined as luminance ratio

$C_1 = L_2/L_1$ (usually for successive stimuli),

or by the following formula

$C_2 = (L_2 - L_1)/L_1$ (usually for surfaces viewed simultaneously);

when the areas of different luminance are comparable in size and it is desired to take an average, the following formula may be used instead

$C_3 = (L_2 - L_1)/0.5(L_2 + L_1)$

where L_1 is the luminance of the background, or largest part of the visual field; and L_2 is the luminance of the object.

Luminance meter (*as per CIE 17.4:1987; 845-05-17*)
Instrument for measuring luminance.

Luminance uniformity (*as per BS EN 12665:2002*)
Ratio of minimum luminance to average luminance.

Note: Use is also made of the ratio of minimum luminance to maximum luminance, in which case this should be specified explicitly.

Luminous area
The area of a lamp or luminaire that emits light. For a flat surface, the projected area varies with the cosine of the angle between the direction of view and the normal to the surface. For a spherical surface, the projected area is constant for all directions of view. For less regular solids, e.g. a

surface diffuser luminaire, the luminous surfaces in various planes must be calculated separately. For specular reflectors or prismatic optics, the 'flashed luminous area' varies with viewing angle and no simple relationship applies in practice.

Luminous efficacy of radiation (K) *(as per CIE 17.4:1987; 845-01-56)*
Quotient of the luminous flux Φ_v by the corresponding radiant flux Φ_e.

$$K = \frac{\Phi_v}{\Phi_e}$$

Unit: lm/W.

Note

When applied to monochromatic radiations, the maximum value of $K(\lambda)$ is denoted by the symbol K_m.
$K_m = 683 \, \text{lm/W}$ for $\nu_m = 540 \times 10^{12} \, \text{Hz}$ ($\lambda_m \approx 555 \, \text{nm}$) for photopic vision.
$K'_m = 1700 \, \text{lm/W}$ for $\lambda'_m = 507 \, \text{nm}$ for scotopic vision.
For other wavelengths: $K(\lambda) = K_m V(\lambda)$ and $K'(\lambda) = K'_m V'(\lambda)$

Luminous efficacy of a source (η_v:η) *(as per CIE 17.4:1987; 845-01-55)*
Quotient of the luminous flux emitted by the power consumed by the source.

Unit: lm/W.

Notes: It must be specified whether or not the power dissipated by auxiliary equipment such as ballasts, etc., if any, is included in the power consumed by the source.

Luminous environment *(as per BS EN 12665:2002)*
Lighting considered in relation to its physiological and psychological effects. (*See also* CIE 17.4:1987; 845-09-03.)

Luminous flux *(as per CIE 17.4:1987; 845-01-25)*
Quantity derived from radiant flux Φ_e by evaluating the radiation according to its action upon the CIE standard photometric observer. For photopic vision

$$\Phi_v = K_m \int_0^\infty \frac{d\Phi_e(\lambda)}{d\lambda} \times V(\lambda)d\lambda$$

where $\frac{d\Phi_e(\lambda)}{d\lambda}$ is the spectral distribution of the radiant flux; and $V(\lambda)$ is the spectral luminous efficiency.

Unit: lumen (lm).

Note: For the values of K_m (photopic vision) and K'_m (Scotopic vision) (*See CIE 17.4:1987; 845-01-56 Luminous efficacy of radiation.*)

Luminous intensity (of a point source in a given direction) (*I*) (*as per CIE 17.4:1987; 845-01-31*)
Quotient of the luminous flux $d\Phi$ leaving the source and propagated in the element solid angle $d\Omega$ containing the given direction, by the element solid angle.

$$I = \frac{d\Phi}{d\Omega}$$

Unit: candela.

Luminous intensity distribution
The spatial distribution of the luminous intensity of a lamp or luminaire in all directions. Luminous intensity distributions are often shown in the form of polar and cartesian diagrams, in specified planes usually vertical or horizontal, or a table (*I* table) and may be expressed in candelas per 1000 lamp lumens.

Lux (lx) (*as per CIE 17.4:1987; 845-01-52*)
SI unit of illuminance: Illuminance produced on a surface area $1\,m^2$ by a luminous flux of $1\,lm$ uniformly distributed over that surface.

Unit: lx.

Maintained level (of average road surface luminance, average or minimum illuminance on road area, average hemispherical Illuminance, minimum semi-cylindrical illuminance or minimum vertical plane illuminance) (*as per EN 13201-2:2003*)
Design level reduced by a maintenance factor to allow for depreciation.

Maintained illuminance (E_m) (*as per BS EN 12665:2002*)
Value below which the average illuminance on the specified area should not fall. It is the average illuminance at the time maintenance should be carried out.

Unit: $lx = lm/m^2$.

Maintained luminance (L_m) (*as per BS EN 12665:2002*)
Value below which average luminance should not fall. It is the average luminance at the time maintenance should be carried out.

Unit: $cd\,m^{-2}$.

Maintenance cycle (*as per BS EN 12665:2002*)
Repetition of lamp replacement, lamp/luminaire cleaning and room surface cleaning (if applicable) intervals.

Maintenance factor
See Light loss factor.

Maintenance schedule (*as per BS EN 12665:2002*)
Set of instructions specifying maintenance cycle and servicing procedures.

Matrix signal (*as per CIE 17.4:1987; 845-11-04*)
Sign designed to display a variable message by means of an array of elementary units, each of which can be individually illuminated or otherwise altered in appearance.

Maximum illuminance (E_{max}) (*as per BS EN 12665:2002*)
Highest illuminance at any relevant point on the specified surface.

Unit: $lx = lm/m^2$.

Note: The relevant points at which the illuminances are determined shall be specified in the appropriate application standard.

Maximum luminance (L_{max}) (*as per BS EN 12665:2002*)
Highest luminance of any relevant point on the specified surface.

Unit: $cd\,m^{-2}$.

Note: The relevant points at which the luminances are determined shall be specified in the appropriate application standard.

Measurement field (of Photometer) (*as per BS EN 12665:2002*)
Area including all points in object space, radiating towards the acceptance area of the detector.

Measurement grid
A series of measurement points set out in a grid formation. Unless an alternative grid is agreed between the parties concerned as being satisfactory, the grid of measurement points should be the same as the calculation grid.

Measurement point
A point on the working plane where a photometric measurement is to be taken.

Mesopic vision
Vision associated with illuminated environments lying in the range .001 to 3 cd – levels found in many exterior lighting environments. It is an ill-defined range lying between photopic and scotopic vision. The cone cells in the eye are on the verge of no longer receiving enough light for them to be stimulated and the rod cells are beginning to take over as the prime receptor. The eye's sensitivity to the bluer end of the spectrum is increasing from that in photopic vision as the level falls. Research with white light sources (with more blue content than, say, high pressure sodium) is concerned with their visual effectiveness in the mesopic range.

Metal halide lamp (*as per CIE 17.4:1987; 845-07-25*)
A high intensity discharge lamp in which the major portion of the light is produced from a mixture of metallic vapour and the product of the dissociation of halides.

Note: The term covers clear and phosphor-coated lamps.

Minimum illuminance (E_{min}) (*as per BS EN 12665:2002*)
Lowest illuminance at any relevant point on the specified surface.

Unit: $lx = lm/m^2$.

Note: The relevant points at which the illuminances are determined shall be specified in the appropriate application standard.

Minimum illuminance (on a road area) (E_{min}) (*as per EN 13201-2:2003*)
Lowest illuminance on a road surface.

Unit: $lux = lx$.

Minimum luminance (L_{min}) (*as per BS EN 12665:2002*)
Lowest luminance of any relevant point on the specified surface.

Unit: $cd\,m^{-2}$.

Note: The relevant points at which the luminances are determined shall be specified in the appropriate application standard.

Mixed traffic
Traffic that consists of motor vehicles, cyclists, pedestrians, etc.

Monochromatic radiation (*as per CIE 17.4:1987; 845-01-07*)
Radiation characterised by a single frequency. In practice, radiation of a very small range of frequencies which can be described by stating a single frequency.

Note: The wavelength in air or in vacuo is also used to characterise a
monochromatic radiation.

Motor traffic (motorised traffic)
Traffic that consists of motorised vehicles only. It depends on national
legislation which vehicle types are included in this classification. In some
countries it includes only vehicles which are capable of maintaining a minimum speed. In others, mopeds are not considered as motorised traffic.

Mounting height (h_m)
The nominal vertical distance between the photometric centre of the luminaire (or array) and the surface of the defined area.

Multi-level road junction
A road junction where at least one road passes over another.

Munsell system
A system of surface colour classification using uniform colour scales of hue,
value and chroma. A typical Munsell designation of colour is 5G3/4, where
5G (green) is the hue reference, 3 is the value and 4 is the chroma reference
number.

Natural sky glow
That part of the sky glow which is attributable to radiation from celestial
sources and luminescent processes in the Earth's upper atmosphere.

Nominal height (*as per BS EN 40-1:1991*)
The distance between the centre line of the point of entry of the luminaire
and the intended ground level, for a column planted in the ground, or the
bottom of the flange plate, for a column with a flange plate.

Non-cut-off luminaire (road lighting)
A classification applied to a luminaire which when installed in a specified
design attitude the maximum permissible value of intensity emitted does
not exceed 1000 cd at 90°.

Obtrusive light
Light, outside the area to be lit, which, because of quantitative, directional
or spectral attributes in a given context, gives rise to annoyance, discomfort,
distraction or a reduction in the ability to see essential information, e.g. at
signal lights.

Opaque medium (*as per CIE 17.4:1987; 845-04-110*)
Medium which transmits no radiation in the spectral range of interest.

Optical light output ratio (of a luminaire) (*as per CIE 17.4:1987; 845-09-38*)
Ratio of the total flux of the luminaire, measured under specified conditions, to the sum of the individual luminous fluxes of the lamps when inside the luminaire.

Note: For luminaires using incandescent lamps only, the optical light output ratio and the light output ratio are the same in practice.

Optical radiations (*as per CIE 17.4:1987; 845-01-02*)
Electromagnetic radiations at wavelengths between the regions of transition to X-rays ($\lambda \approx 1$ nm) and the region of transition to radio waves ($\lambda \approx 1$ mm)

Outreach (road lighting) (*as per BS 5489: Part 1: 1992*)
The horizontal distance for column or wall-mounted luminaires between the centreline of the column at ground level, or wall-bracket mounting surface and the photometric centre of the luminaire.

Overall uniformity ratio of illuminance
Ratio of the minimum illuminance to the average illuminance over the defined area.

Overall uniformity (of road surface luminance, illuminance on a road area or hemispherical illuminance) (U_o) (*as per EN 13201-2:2003*)
Ratio of the lowest to the average value.

Overhang (road lighting) (*as per BS 5489: Part 1: 1992*)
The distance measured horizontally between the photometric centre of a luminaire and the adjacent edge of carriageway. The distance is taken to be positive if the luminaire is in front of the edge and negative if it is behind the edge.

Note: For installation purposes, the centre of a lamp or lamps may be regarded as the photometric centre of a luminaire.

Parting zone (of a tunnel) (*as per BS 5489-2:2003*)
First part of the open road directly after the exit portal of the tunnel.

Note: The parting zone is not a part of the tunnel, but is closely related to the tunnel lighting. The parting zone begins at the exit portal.

Penning mixture
A mixture that replaces a single inert gas in order to reduce the striking voltage of a lamp.

Perceived light (see also Light) (*as per CIE 17.4:1987; 845-02-17*)
Universal and essential attribute of all perceptions and sensations that are peculiar to the visual system.

Note: Light is normally, but not always, perceived as a result of the action of a light stimulus on the visual system.

Photometer (*as per CIE 17.4:1987; 845-05-15*)
Instrument for measuring photometric quantities.

Photometry (*as per CIE 17.4:1987; 845-05-09*)
Measurement of quantities referring to radiation evaluated according to a given spectral luminous efficiency function, e.g. $V(\lambda)$ or $V'(\lambda)$.

Photopic vision
Vision associated with a 'light adapted' eye (*see Adaptation*) at relatively high light levels of typically above $3\,\mathrm{cd\,m^{-2}}$. In this situation the cone cells in the eye are stimulated and vision is in colour.

Planckian locus (*as per CIE 17.4:1987; 845-03-41*)
The locus of points in a chromaticity diagram that represents chromaticities of the radiation of Planckian radiators at different temperatures.

Planting depth (of a lighting column)
The length of the lighting column below the intended ground level.

Pole
An alternative term used for a lighting column. Generally used to imply a lighting column that carries one or more luminaires for sports, effect lighting or floodlighting.

Post mount (road lighting luminaires)
Luminaire-fixing arrangement where there is no bracket and one part of the underside of a luminaire (other than the centre) is mounted directly on top of a column shaft.

Post top (road lighting luminaires)
Luminaire-fixing arrangement where there is no bracket and the centre of the underside of a luminaire is mounted directly on top of a column shaft.

Post-top lighting column
A straight lighting column without bracket to support the luminaire (post-top luminaire) directly.

Power factor
Ratio of the root mean square power in Watts to the product of the root mean square values of voltage and current; for sinusoidal waveforms the power factor is also equal to the cosine of the angle of phase difference between voltage and current.

Projector (*as per CIE 17.4:1987; 845-10-25*)
Luminaire using reflection and/or refraction to increase luminous intensity within a limited solid angle.

Projection
See Bracket projection.

Purkinje phenomenon (*as per CIE 17.4:1987; 845-02-14*)
Reduction in the brightness of a predominantly long-wavelength colour stimulus relative to that of a predominantly short-wavelength colour stimulus when the luminances are reduced in the same proportion from photopic to mesopic or scotopic levels without changing the respective relative spectral distribution of the stimuli involved.

Note: In passing from photopic to mesopic or scotopic vision, the spectral luminous efficiencies change, the wavelength of maximum efficiency being displaced towards the shorter wavelengths.

R table
A table of (reduced) luminance coefficients for a surface that for different directions of incident illuminance and view at a point on the surface allows the luminance at that point to be calculated.

Radiance (in a given direction, at a given point of a real or imaginary surface) (L_e; L) (*as per CIE 17.4:1987; 845-01-34*)
Quantity defined by the formula

$$L_e = \frac{d\Phi_e}{dA \times \cos \Theta \times d\Omega}$$

where $d\Phi_e$ is the radiant flux transmitted by an elementary beam passing through the given point and propagating in the solid angle $d\Omega$ containing the given direction; dA is the area of a section of that beam containing the given point; and Θ is the angle between the normal to that section and the direction of the beam.

Unit: $W\,m^{-2}/sr$.

Radiant exitance (at a point of a surface) (M_e; M) (*as per CIE 17.4:1987; 845-01-47*)
Quotient of the radiant flux $d\Phi_e$ leaving an element of the surface containing the point, by the area dA of that element.

Equivalent definition: Integral, taken over the hemisphere visible from the given point of the expression $L_e \times \cos\Theta \times d\Omega$. Where L_e is the radiance at the given point in the various directions of the emitted elementary beam of solid angle $d\Omega$; and Θ is the angle between any of the beams and the normal to the surface at the given point.

$$M_e = \frac{d\Phi_e}{dA} = \int_{2\pi sr} L_e \cos\Theta \, d\Omega$$

Unit: W/m^2.

Radiant flux; radiant power (Φ_e; Φ; P) (*as per CIE 17.4:1987; 845-01-24*)
Power emitted, transmitted or received in the form of radiation.

Unit: W.

Rating (of a type of lamp) (*as per CIE 17.4:1987; 845-07-58*)
The set of rated values and operating conditions of a lamp which serves to characterise and designate it.

Rated luminous flux (of a type of lamp) (*as per CIE 17.4:1987; 845-07-59*)
The value of the initial luminous flux of a given type of lamp declared by the manufacturer or the responsible vendor, the lamp being operated under specified conditions.

Unit: lumens (lm).

Notes

1 The initial luminous flux is the luminous flux of a lamp after a short ageing period as specified in the relevant lamp standard.
2 The rated luminous flux is sometimes marked on the lamp.

Rated power (of a type of lamp) (*as per CIE 17.4:1987; 845-07-60*)
The value of the power of a given type of lamp declared by the manufacturer or the responsible vendor, the lamp being operated under specified conditions.

Unit: W.

Note: The rated power is usually marked on the lamp.

Reference ballast (*as per BS EN 12665:2002*)
Special inductive-type ballast designed for the purpose of providing comparison standards for use in testing ballasts, for the selection of reference lamps and for testing regular production lamps under standardised conditions. (*See also CIE 17.4:1987; 845-08-36.*)

Reference lamp (*as per BS EN 12665:200*)
Discharge lamp selected for the purpose of testing ballasts and which, when associated with a reference ballast under specified conditions, has electrical values which are close to the objective values given in a relevant specification. (*See also CIE 17.4:1987; 845-07-55.*)

Reference surface (*as per BS EN 12665:2002*)
Surface on which illuminance is measured or specified. (*See also CIE 17.4:1987; 845-09-49.*)

Reflectance (for incident radiation of given spectral composition, polarisation and geometrical distribution) (ρ)
Ratio of reflected radiant or luminous flux to the incident flux in the given condition. (*See also CIE 17.4:1987; 845-04-58.*)

Reflectance (factor) (R, ρ)
The ratio of the luminance flux reflected from a surface to the luminous flux incident on it. Except for matt surfaces, reflectance depends on how the surface is illuminated but especially on the direction of the incident light and its spectral distribution. The value is always less than unity and is expressed as either a decimal or as a percentage.

Reflection (*as per CIE 17.4:1987; 845-04-42*)
Process by which radiation is returned by a surface or a medium, without change of frequency of its monochromatic components.

Notes

1   Part of the radiation falling on a medium is reflected at the surface of the medium (surface reflection); another part may be scattered back from the interior of the medium (volume reflection).
2   The frequency is unchanged only if there is no Doppler effect due to the motion of the materials from which the radiation is returned.

Reflector (*as per CIE 17.4:1987; 845-10-33*)
Device used to alter the spatial distribution of the luminous flux from a source and depending on the phenomenon of reflection.

Refraction *(as per CIE 17.4:1987; 845-04-100)*
Process by which the direction of a radiation is changed as a result of changes in its velocity of propagation in passing through an optically non-homogeneous medium, or in crossing a surface separating different media.

Refractor *(as per CIE 17.4:1987; 845-10-32)*
Device used to alter the spatial distribution of the luminous flux from a source and depending on the phenomenon of refraction.

Regular reflection; specular reflection *(as per CIE 17.4:1987; 845-04-45)*
Reflection in accordance with the laws of geometrical optics, without diffusion.

Relative spectral sensitivity (curve) $V(\lambda)$
The visual effect of light energy at a particular wavelength is not uniform across the visible spectrum. The $V(\lambda)$ curve shows the relative effect of light at each wavelength, with that at 560 nm being equated to unity (ISO/IEC 10527). *(See also CIE 17.4:1987; 845-05-57.)*

Retroreflection *(as per CIE 17.4:1987; 845-04-92)*
Reflection in which radiation is returned in directions close to the opposite of the direction from which it came, this property being maintained over wide variations of the direction of the incident rays.

Retroreflector *(as per CIE 17.4:1987; 845-04-93)*
A surface or device from which most of the reflected radiation is retrore-flected.

Retinal illuminance
See Troland.

Roundabout
Road junction that has one-way movement of traffic round a traffic island.

Safety lighting *(as per CIE 17.4:1987; 845-09-12)*
That part of Emergency lighting provided to ensure the safety of people involved in a potentially hazardous process.

Scattering
See Diffusion.

Scotopic vision
Vision associated with a 'dark adapted' eye *(see Adaptation)* at very low light levels – typically in the range below $0.001\,\mathrm{cd\,m^{-2}}$. In this situation only the rod cells are stimulated and vision is monochromatic and, as the rod cells cover the entire retina outside of the fovea, is associated with peripheral vision.

Searchlight *(as per CIE 17.4:1987; 845-10-26)*
A high intensity projector having an aperture usually greater than $0.2\,\mathrm{m}$ and giving an approximately parallel beam of light.

Semi-cut-off luminaire (road lighting)
A classification applied to a luminaire which when installed in a specified design attitude the maximum permissible value of intensity emitted does not exceed 50 cd/1000 lm at 90° (up to a maximum of 1000 cd whatever the luminous flux emitted) and 100 cd/1000 lm at 80°.

Semi-cylindrical illuminance (at a point) (E_{sc}) *(as per BS EN 12665:2002)*
Total luminous flux falling on the curved surface of a very small semi-cylinder located at the specified point, divided by the curved surface area of the semi-cylinder. The axis of the semi-cylinder is taken to be vertical unless stated otherwise. The direction of the curved surface should be specified.

Unit: $\mathrm{lx} = \mathrm{lm/m^2}$.

Semi-direct lighting *(as per BS EN 12665:2002)*
Lighting by means of luminaires having a distribution of luminous intensity such that the fraction of the emitted luminous flux directly reaching the working plane, assumed to be unbounded, is 60–90 per cent. *(See also CIE 17.4:1987; 845-09-15.)*

Semi-indirect lighting *(as per BS EN 12665:2002)*
Lighting by means of luminaires having a distribution of luminous intensity such that the fraction of the emitted luminous flux directly reaching the working plane, assumed to be unbounded, is 10–40 per cent. *(See also CIE 17.4:1987; 845-09-17.)*

Service strip (road lighting)
A strip of land adopted by the Highway Authority running alongside and adjacent to a carriageway (usually a shared access way) and used by the utility companies for their apparatus. The surface of the service strip is usually maintained by the adjacent property owner.

Set back (road lighting)
The shortest distance from the forward face of a column to the edge of a carriageway.

Shade (*as per CIE 17.4:1987; 845-10-33*)
Screen which may be made of opaque or diffusing material and which is designed to prevent a lamp from being directly visible.

Short-arc lamp (*as per CIE 17.4:1987; 845-07-34*)
An arc lamp, generally of very high pressure, in which the distance between the electrodes is of the order of 1–10 mm.

Note: Certain mercury vapour or xenon lamps belong to this type.

Side entry (road lighting luminaires)
A luminaire-fixing arrangement where provision is made to attach the luminaire to a substantially horizontal bracket (0 to +15°).

Sign post
Support intended to hold one or more traffic signs and associated lanterns, consisting of one or more parts: a post, possibly an extension piece and, if necessary, a bracket.

Silhouette vision
A process whereby information within the three-dimensional visual scene is seen as dark against a lighter background, i.e. negative contrast.

Sky glow
The brightening of the night sky that results from the reflection of radiation (visible and non-visible), scattered from the constituents of the atmosphere (gaseous molecules, aerosols and particulate matter), in the direction of observation. (*See also Natural sky glow and Artificial sky glow.*)

Skylight (*as per CIE 17.4:1987; 845-09-83*)
Visible part of diffuse sky radiation.

Note: When dealing with actinic effects of optical radiations, this term is commonly used for radiations extending beyond the visible region of the spectrum.

Small car park
A car park restricted to the use of a single company, e.g. office buildings and industrial plants or a public car park with a capacity of fewer than 100 vehicles for district shopping centres, sports centres or multipurpose building complexes.

Solar radiation (*as per CIE 17.4:1987; 845-09-76*)
Electromagnetic radiation from the Sun.

Solid angle
The angle subtended by an area at a point and equal to the quotient of the projected area on a sphere centred on the point, by the square of the radius of the sphere; expressed in steradians.

Unit: Steradian (sr).

Spacing (in an installation) (*as per BS EN 12665:2002*)
Distance between the light centres of adjacent luminaires of the installation. (*See also CIE 17.4:1987; 845-09-66.*)

Spacing to height ratio (*as per BS EN 12665:2002*)
Ratio of spacing to the height of the geometric centres of the luminaires above the reference plane.

Notes

1   For indoor lighting the reference plane is usually the horizontal working plane; for exterior lighting the reference plane is usually the ground.
2   For road lighting the spacing to height ratio is the distance between consecutive columns or masts divided by the mounting height.

(Spatial) distribution of luminous intensity (of a source) (*as per CIE 17.4:1987; 845-09-24*)
Display, by means of curves or tables, of the value of the luminous intensity of the source as a function of direction in space.

Specified task
The activity for which the lighting design has been prepared.

Spectral (power or energy) distribution
Sometimes referred to as 'spectral composition'. The variation of radiant power (or energy) over a range of wavelengths.

Spectral reflection
Reflection without diffusion in accordance with the laws of optical reflection as in a mirror.

Spectrum (*as per CIE 17.4:1987; 845-01-08*)
Display or specification of the monochromatic components of the radiation considered.

Notes

1   There are line spectra, continuous spectra and spectra exhibiting both these characteristics.
2   This term is also used for spectral efficiencies (excitation spectrum, action spectrum).

Specular factor (S_1) (of a road surface)
The specular reflection of a road surface, defined as

$$S_1 = \frac{R(0, 2)}{R(0, 0)}$$

where r is the reflection value at specific angles of β and γ, i.e. for $R(0, 2)$ then $\beta = 0°$, $\gamma = \arctan 2 = 63.4°$, and for $R(0, 0)$ $\beta = 0°$, $\gamma = \arctan 0$.

Specular reflection
See Regular reflection.

Speed limit
The maximum legally allowed speed.

Spherical illuminance (at a point) (E_o)
Total luminous flux falling onto the whole surface of a very small sphere located at the specified point divided by the total surface area of the sphere. (*See also CIE 17.4:1987; 845-01-40 spherical irradiance.*)

Unit: $lx = lm/m^2$.

Technically defined: Quantity defined by the formula

$$E_o = \int_{4\pi sr} L d\Omega$$

where $d\Omega$ is the solid angle of each elementary beam passing through the given point; and L its luminance at that point.

Spill light (stray light)
Light emitted by a lighting installation, which falls outside the boundaries of the property on which the installation is sited.

Spot lamp renewal
See ad-hoc lamp renewal.

Spotlight *(as per CIE 17.4:1987; 845-10-27)*
A projector having an aperture usually smaller than 0.2 m and giving a concentrated beam of light of usually not more than 0.35 radians (20°) divergence.

Spotlighting *(as per BS EN 12665:2002)*
Lighting designed to increase considerably the illuminance of a limited area or of an object relative to the surroundings, with minimum diffused lighting. *(See also CIE 17.4:1987; 845-09-22.)*

Standby lighting *(as per CIE 17.4:1987; 845-09-13)*
That part of Emergency lighting provided to enable normal activities to continue substantially unchanged.

Starter *(as per CIE 17.4:1987; 845-08-32)*
A starting device, usually for fluorescent lamps, which provides for the necessary pre-heating of the electrodes and, in combination with the series impedance of the ballast, causes a surge in the voltage applied to the lamp.

Starting device *(as per CIE 17.4:1987; 845-08-31)*
Apparatus which provides, by itself or in combination with other components in the circuit, the appropriate electrical conditions needed to start a discharge lamp.

Steradian (sr)
SI unit of solid angle: Solid angle that, having its vertex at the centre of a sphere, cuts off an area of the surface of the sphere equal to that of a square with sides of length equal to the radius of the sphere. A complete sphere subtends 4πsr from the centre. *(See also CIE 17.4:1987; 845-01-20.)*

Stiles–Crawford effect (of the first kind); (directional effect) (see also CIE 17.4:1987; 845-02-15)
Decrease of the brightness of a light stimulus with increasing eccentricity of the position of entry of the light through the pupil.

Note: If the variation is in hue and saturation instead of in brightness, the effect is called Stiles–Crawford effect of the second kind.

Stopping distance (of a tunnel) *(as per BS 5489-2:2003)*
Distance needed to bring a vehicle, driving at design speed, to a complete standstill.

Stroboscopic effect (*as per BS EN 12665:2001*)
Apparent change of motion and/or appearance of a moving object when the object is illuminated by a light of varying intensity.

Note: To obtain apparent immobilisation or constant change of movement, it is necessary that both the object movement and the light intensity variation are periodic, and some specific relation between the object movement and light variation frequencies exists. The effect is only observable if the amplitude of the light variation is above certain limits. The motion of the object may be rotational or translational.

Sunlight (*as per CIE 17.4:1987; 845-09-82*)
Visible part of direct solar radiation.

Note: When dealing with actinic effects of optical radiations, this term is commonly used for radiations extending beyond the visible region of the spectrum.

Surround ratio (of illumination of a carriageway of a road) (SR) (*as per EN 13201-2:2003*)
Average illuminance on strips just outside the edges of the carriageway in proportion to the average illuminance on strips just inside the edges.

Threshold increment (TI) (*as per EN 13201-2:2003*)
Measure of the loss of visibility caused by the disability glare of the luminaires of a road lighting installation.

Threshold zone (of a tunnel) (*as per BS 5489-2:2003*)
First part of the tunnel, directly after the portal.

Note: The threshold zone begins at the entrance portal.

Top entry (road lighting luminaires)
A luminaire-fixing arrangement in which the luminaire is attached to the supporting structure by a connection on its top surface.

Traffic bollard (*as per CIE 17.4:1987; 845-11-64*)
Post used to indicate an obstruction or to regulate traffic. It may be internally illuminated and may incorporate a regulatory traffic sign. (*See also Transilluminated traffic bollard.*)

Traffic flow
The number of vehicles passing a specific point in a stated time in stated direction(s).

Note: Average daily traffic (ADT, vehicles per day) is the most used concept in traffic planning and is generally known. Peak hour traffic (vehicles per hour) is on rural roads 10 per cent and in urban areas 12 per cent of ADT. On undivided roads, number of vehicles per hour per lane can be calculated by dividing peak hour value by the total number of lanes. If the actual directional distribution is not known on dual carriageway roads, assumption of 1:2 can be made. Then the higher flow will be divided by the number of lanes of this carriageway.

Traffic lane
A strip of carriageway intended to accommodate a single line of moving vehicles.

Transilluminated traffic bollard (TTB) (*as per BS EN 12899-2:2001*)
A totally or partially transilluminated device placed within the highway to warn drivers of obstructions.
Type 1 TTB: a TTB which may incorporate one or more traffic signs or plain areas as alternatives.
Type 2 TTB: a TTB which may support one or more traffic signs.

Transition zone (of a tunnel) (*as per BS 5489-2:2003*)
Part of a tunnel following directly after the threshold zone.

Note: The transition zone stretches from the end of the threshold zone to the beginning of the interior zone. In the transition zone, the lighting level is decreased from the level at the end of the threshold zone to the level of the interior zone.

Transmission (*as per CIE 17.4:1987; 845-04-43*)
Passage of radiation through a medium without change of frequency of its monochromatic parts.

Transmittance (for incident radiation of a given spectral composition, polarisation and geometrical distribution) (τ)
Ratio of the transmitted radiant or luminous flux to the luminous flux incident in the given conditions. (*See also CIE 17.4:1987; 845-04-59.*)

Translucent medium (*as per CIE 17.4:1987; 845-04-109*)
Medium which transmits visible radiation largely by diffuse transmission, so that objects are not seen distinctly through it.

Transparent medium (*as per CIE 17.4:1987; 845-04-108*)
Medium in which the transmission is mainly regular and which usually has a high regular transmittance in the spectral range of interest.

Note: Objects may be seen distinctly through a medium which is transparent in the visible region, if the geometric form of the medium is suitable.

Tristimulus values (of a colour stimulus) (*as per BS EN 12665:2002*)
Amounts of the three reference colour stimuli, in a given trichromatic system, required to match the colour of the stimulus considered. (*See also CIE 17.4:1987; 845-03-22.*)

Note: In the CIE standard colorimetric systems, the tristimulus values are represented by the symbols X, Y, Z and X_{10}, Y_{10}, Z_{10}.

Troland (T_d) (*as per CIE 17.4:1987; 845-02-16*)
Unit used to express a quantity proportional to retinal illuminance produced by a light stimulus. When the eye is viewing a surface of uniform stimulus, the number of trolands is equal to the product of the area in square millimetres of the limiting pupil, natural and artificial, by the luminance of the surface in candelas per square metre.

Note: In computing effective retinal illuminance, absorption, scattering, and reflection losses and the dimension of the particular eye under consideration must be taken into account as well as the Stiles–Crawford effect.

Tungsten halogen lamp (*as per CIE 17.4:1987; 845-07-10*)
Gas-filled lamp containing halogens or halogen compounds, the filament being of tungsten.

Note: Iodine lamps belong to this category.

Uniformity ratio of illuminance (*as per CIE 17.4:1987; 845-09-58*)
Ratio of the minimum illuminance to the average illuminance on the plane.

Uniformity ratio of luminance
Ratio of the minimum luminance to the average luminance over a specified surface.

Unified glare rating (UGR) (*as per CIE 117:1995*)
The measure of discomfort under interior lighting defined by the following formula:

$$\text{UGR} = 8 \log_{10} \left[\frac{0.25}{L_b} \sum \frac{L^2 \omega}{P^2} \right]$$

where L_b is the background luminance (cd m^{-2}); L is the luminance of the luminous parts of each luminaire in the direction of the observer's eye

$(\text{cd}\,\text{m}^{-2})$; ω is the solid angle subtended by the luminous parts of each luminaire at the observer's eye; and P is the Guth Position Index for each luminaire (a function of the displacement of the luminaire from the line of sight of the observer).

Upward light output ratio (of a luminaire) (ULOR) *(as per BS EN 12665:2002)*
Ratio of the upward flux of the luminaire, measured under specified practical conditions with its own lamps and equipment, to the sum of the individual luminous fluxes of the same lamps when operated outside the luminaire with the same equipment, under specified conditions.

Note: For luminaires using incandescent lamps only, the optical light output ratio and the light output ratio are the same in practice.

Upward light ratio (ULR)
Proportion of the flux of a luminaire and/or installation that is emitted at and above the horizontal when the luminaire(s) is mounted in its installed position.

Utilisation factor (of an installation, for a reference surface) *(as per BS EN 12665:2002)*
Ratio of the luminous flux received by the reference surface to the sum of the individual fluxes of the lamps of the installation. *(See also CIE 17.4:1987; 845-09-53.)*

$V(\lambda)$ correction *(as per BS EN 12665:2002)*
Correction of the spectral responsivity of a detector to match the photopic spectral sensitivity of the human eye. *(See also CIE 17.4:1987; 845-01-22 and 845-01-23.)*

$V(\lambda)$ function
See Relative spectral sensitivity.

Vacuum (incandescent) lamp *(as per CIE 17.4:1987; 845-07-08)*
Incandescent lamp in which the luminous element operates in an evacuated bulb.

Value
In the Munsell system, an index of the lightness of a surface ranging from 0 (black) to 10 (white). Approximately related to percentage reflectance by the relationship

$$R = \frac{V(V-1)}{100}$$

where R is reflectance (per cent); and V is value.

Veiling reflections (*as per BS EN 12665:2002*)
Specular reflections that appear on the object viewed and that partially or wholly obscure the details by reducing contrast. (*See also CIE 17.4:1987; 845-02-55.*)

Vertical illuminance
The illuminance on a vertical plane at a specified point.

Visible radiation (*as per CIE 17.4:1987; 845-01-03*)
Any optical radiation capable of causing a visual sensation directly.

Note: There are no precise limits for the spectral range of visible radiation since they depend upon the amount of radiation power reaching the retina and the responsivity of the observer. The lower limit is generally taken between 360 and 400 nm and the upper limit between 760 and 830 nm.

Visual acuity (*as per BS EN 12665:2002*)
Qualitatively: Capacity for seeing distinctly fine details that have very small angular separation.

Quantitatively: Any of a number of measures of spatial discrimination such as the reciprocal of the value of the angular separation in minutes of arc of two neighbouring objects (points or lines or other specified stimuli) which the observer can just perceive to be separate. (*See also CIE 17.4:1987; 845-02-47.*)

Visual comfort (*as per BS EN 12665:2002*)
Subjective condition of visual well-being induced by the visual environment.

Visual disability
Any restriction or lack (resulting from a visual impairment) of ability to perform an activity in the manner or within the range considered normal for a human being.

Visual environment
The environment as seen by an observer.

Visual field (*as per BS EN 12665:2002*)
Area or extent of physical space visible to an eye at a given position and direction of view.

Note: It should be stated whether the visual field is monocular or binocular.

Visual handicap
A visual disadvantage for a given individual, resulting from an impairment or a disability, that limits or prevents the fulfilment of a role that is normal for that individual. This may be due to the arrangement of lighting or task and would apply to any individual in the same situation. It may also be peculiar to the individual. Where this is a physical disability, the individual is said to have 'low vision' and may benefit from special lighting or some optical aid.

Visual impairment
Any loss or abnormality of vision.

Visual performance (*as per BS EN 12665:2002*)
Performance of the visual system as measured, for instance, by the speed and accuracy with which a visual task is performed. (*See also CIE 17.4:1987; 845-09-04.*)

Visual guidance
The optical and geometrical means that ensure that motorists are given adequate information on the course of the road.

Weight
An approximate correlation of Munsell value modified to give in conjunction with greyness, subjective equality of brightness in the various hues (from BS 5252).

Windage area
The projected area of the luminaire, in m^2, on the vertical plane normal to the direction of the wind.

Working plane
The horizontal, vertical, or inclined plane in which the visual task lies. If no information is available, the working plane may be considered to be horizontal and at ground level.

Notes

Foreword

1 Copies available at: www.homeoffice.gov.uk/rds/horspubs1.html.

2 Social and environmental elements

1 Pease, K., *Lighting and Crime*, Huddersfield University (1998).
2 A copy of the report summary is available for download from www.ile.org.uk.
3 Farrington, David and Welsh, Brandon, 'Effects of Improved Street Lighting on Crime: A Systematic Review', HORS 251, Home Office, London, 2002.
4 Farrington, David and Walsh, Brandon, 'Effects of Improved Street Lighting on Crime: A Systematic Review', HORS 251, Home Office, London, 2002, p. 39.
5 Pease, K., *Lighting and Crime*, Huddersfield University (1998).
6 Farrington, David and Welsh, Brandon, 'Effects of Improved Street Lighting on Crime: A Systematic Review', HORS 251, Home Office, London, 2002.
7 Farrington, David and Welsh, Brandon, 'Effects of Closed Circuit Television on Crime: A Systematic Review', HORS 252, Home Office, London, 2002.
8 Farrington, David and Welsh, Brandon, 'Effects of Closed Circuit Television on Crime: A Systematic Review', HORS 252, Home Office, London, 2002.
9 Painter, K. and Farrington, D.P., 'The financial benefits of improved street lighting, based on crime reduction'. *Lighting Research and Technology*, 33, 3–12 (2001).
10 Graham Phoenix, personal interview, 22 August 2001.
11 Steven Power, personal interview, 4 September 2001.
12 SustainaLite, www.sustainalite.co.uk.
13 House of Commons Environment Committee, Sustainable Waste Management Inquiry, 1998.
14 Williams, P., *Waste Treatment and Disposal*, 1998.
15 Our Common Future (The Brundtland Report), Report of the 1987 World commission on Environment and Development.
16 UK Department of the Environment, Transport and the Regions, 'A Better Quality of Life: A Strategy for Sustainable Development for the United Kingdom', 1999.
17 UK Department of the Environment and Welsh Office, 'Making Waste Work: A Strategy for Sustainable Waste Management in England and Wales', 1995.

3 Equipment

1 For the majority of outdoor lighting installations, the user is the community.
2 There is some evidence that manufacturers' laboratories are not good at reproducing outdoor conditions.

4 Techniques for particular applications

1 Loe, D., Mansfield, K. and Rowlands, T., 'Appearance of the lit environment and its relevance in lighting design: Experimental study', *Lighting Research and Technology,* 26 March 1994.
2 Tregenza, P. and Loe, D., *The Design of Lighting* (Chapter 12, 'The exterior of buildings'), 1998.
3 Lewin, Dr I., 'Lamp spectral effects at roadway lighting levels', *Lighting Journal,* Vol. 64/2, March/April 1999.
4 Lewin, Dr I., 'Visibility factors in outdoor lighting design (Parts 1 and 2)', *Lighting Journal,* Vol. 64/6, November/December 1999 and Vol. 65/1, January/February 2000.
5 Sorenson.
6 Department of Transport, Advice Note TA 49/86 'Appraisal of New and Replacement Lighting on Trunk Roads and Trunk Road Motorways', Crown Office, 1986 and Scottish Office Industry Department, Technical Memorandum SH6/89, 'Appraisal of New and Replacement Lighting on Trunk Roads and Trunk Road Motorways', Crown Office, 1989.
7 Farrington, David and Walsh, Brandon, 'Effects of Improved Street Lighting on Crime: A Systematic Review', HORS 251, Home Office, London, 2002.
8 Outdoor only.

Appendix

1 C.I.E. No. 19.
2 Adrian and Eberbach, 1968/1969.
3 Fisher and Christie, 1965.
4 Economopoulos.
5 Narisada.
6 Waldram.
7 Gallagher and Meguire.

Bibliography

Books

Boer, J.B. de, *Public Lighting*, Philips Tech. Lib., 1967.

Brown, C.N., *JW Swan and the Invention of the Incandescent Lamp*, Science Museum, 1978.

Constant, M., *The Principles and Practice of CCTV*, 2nd edn, Paramount Publishing, ISBN 0-947665-20-X.

Gardner, C. and Molony, R., *Light: Re-interpreting Architecture*, Rotovision, 2001.

Hays, D., *Light on the Subject,* Limelight Edition, 1989.

Henderson, S.T. and Marsden, A.M., *Lamps and Lighting*, Edward Arnold, 1972, ISBN 0 7131 3267 1.

Jenkins, F.A. and White, H.A., *Fundamentals of Optics*, McGraw-Hill, 1957.

Lyon, K.W. and Scott, D.W., *Light*, Edward Arnold Publishers, 1960.

Philips, D., *Lighting Historic Buildings*, Architectural Press, 1997.

Pritchard, D.C., *Lighting*, 5th edn, Longmans, 1995.

Screuder, D.A., *Road Lighting for Safety*, Thomas Telford, 1998, ISBN 0 7277 2616 1.

Van Bommel, W.J.M. and Boer, J.B. de, *Road Lighting*, Kluwer Techniche Boeken, 1980.

Williams, P., *Waste Treatment and Disposal*, John Wiley & Sons, Chichester, 1998.

Waldram, J.M., *Street Lighting*, Arnold, 1952.

Waldram, J.M., *A Manual of Perspective for Lighting Engineers*, CIBSE-L.R. and Tech., 1982.

Papers

Crime

Allatt, P., 'Residential security: Containment and displacement of burglary'. *Howard Journal of Criminal Justice*, 23, 99–116 (1984).

Armitage, R., Smyth, G. and Pease, K., 'The effects of CCTV surveillance in Burnley town centre'. In K. Painter and N. Tilley (eds) *Crime Prevention Studies X*. Monsey, NY: Criminal Justice Press (1999).

Atkins, S., Hosain, S. and Storey, A., 'The influence of street lighting on crime and fear of crime'. Crime Prevention Unit Paper 28. London: Home Office (1991).

Bachner, J.P., 'Effective security lighting'. *Journal of Security Administration*, 9, 59–67 (1985).

Baldrey, P. and Painter, K., 'Watching them watching us'. *The Surveyor*, 30 April, 14–16 (1998).

Barker, M., Geraghty, J. and Webb, B., 'The prevention of street robbery'. Crime Prevention Unit Paper 44. London: Home Office (1993).

Barr, R. and Pease, K., 'Crime placement, displacement and deflection'. In M. Tonry and N. Morris (eds) *Crime and Justice*, Vol. 12 (pp. 277–318). Chicago, Illinois: University of Chicago Press (1990).

Bennett, T. and Gelsthorpe, L., 'Public attitudes towards CCTV in public places'. *Studies on Crime and Crime Prevention*, 5, 72–90 (1996).

Carr, K. and Spring, G., 'Public transport safety: A community right and a communal responsibility'. In R.V. Clarke (ed.) *Crime Prevention Studies 1*. Monsey, NY: Criminal Justice Press (1993).

Clarke, R.V., *Situational Crime Prevention*, 2nd edn. Guilderland, NY: Harrow and Heston (1997).

Davidson, N. and Goodey, J., 'Final report of the Hull street lighting and crime project'. Hull: School of Geography and Earth Sciences, University of Hull (1991).

Demuth, C., *Community Safety in Brighton*. Brighton: Brighton Borough Council Police and Public Safety Unit (1989).

Ditton, J. and McNair, D.G., 'Public enlightenment'. *The Surveyor*, Vol. 181, No. 5282, pp. 18–19 (1994).

Ditton, J. and Nair, G., 'Throwing light on crime: A case study of the relationship between street lighting and crime prevention'. *Security Journal*, Vol. 5, No. 3, July, pp. 125–132 (1994).

Ditton, J. and Nair, G., 'SOX vs SON: Is there really any evidence that white lights are any better than orange ones at preventing crime?'. *The Lighting Journal*, Vol. 60, No. 2, April/May, pp. 91–94 (1995).

Ditton, J., Nair, G. and Bannister, J., 'The cost-effectiveness of improved street lighting as a crime prevention measure'. *The Lighting Journal*, Vol. 61, No. 4, August/September, pp. 251–256 (1996).

Eck, J., 'Preventing crime at places'. In L. Sherman, D.C. Gottfredson, D.L. Mackenzie, J.E. Eck, P. Reuter and S.D. Bushway (eds) *Preventing Crime: What Works, What Doesn't, What's Promising*. Report to US Congress. Washington, DC: NIJ (1997).

Ellingworth, D. and Pease, K., 'Movers and breakers'. *International Journal of Risk, Security and Crime Prevention*, Vol. 3, pp. 35–42 (1998).

Farrington, D.P. and Welsh, B.C., 'Improved street lighting and crime prevention'. *Justice Quarterly*, Vol. 19, No. 2, Academy of Criminal Justice Sciences (2002).

Farrington, D.P. and Welsh, B.C., 'Effects of improved street lighting on crime: A systematic review', Home Office Research Study 251 (2002).

Fennelly, L.J., 'Security surveys'. In L.J. Fennelly (ed.) *Handbook of Loss Prevention and Crime Prevention*, 3rd edn. London: Butterworth (1996).

Fleming, R. and Burrows, J., 'The case for lighting as a means of preventing crime'. *Home Office Research Bulletin*, 22, 14–17 (1986).

Girard, C.M., 'Security lighting'. In L.J. Fennelly (ed.) *Handbook of Loss Prevention and Crime Prevention*, 3rd edn. London: Butterworth (1996).

Griswold, D.B., 'Crime prevention and commercial burglary: A time series analysis'. *Journal of Criminal Justice*, 12, 493–501 (1984).

Griswold, D.B., 'Crime prevention and commercial burglary: A time series analysis'. In R.V. Clarke (ed.) *Situational Crime Prevention: Successful Case Studies*. New York: Harrow and Heston (1992).

Harrisburg, Police Department, 'Final evaluation report of the "High Intensity Street Lighting Program"'.

Harrisburg, Pennsylvania: Planning and Research Section, Staff and Technical Services Division, Harrisburg Police Department (1976).

Hylton, J.B., *Safe schools: A security and loss prevention plan*. London: Butterworth (1996).

Inskeep, N.R. and Goff, C., *A Preliminary Evaluation of the Portland Lighting Project*. Salem, Oregon: Oregon Law Enforcement Council (1974).

Kelling, G.L. and Coles, C.M., *Fixing Broken Windows*. New York: Free Press (1996).

Koch, B.C.M., *The Politics of Crime Prevention*. Aldershot: Ashgate (1998).

Krause, P.B., 'The impact of high intensity street lighting on night-time business burglary'. *Human Factors*, 19, 235–239 (1997).

Lewis, E.B. and Sullivan, T.T., 'Combating crime and citizen attitudes: A study of the corresponding reality'. *Journal of Criminal Justice*, 7, 71–79 (1979).

Lyons, S.L., *Security of Premises: A Manual for Managers*. London: Butterworth (1988).

Mayhew, P.M., 'Crime in public view: Surveillance and crime prevention'. In P.J. Brantingham and P.L. Brantingham (eds) *Environmental Criminology*. Prospect Heights Il: Waveland (1981).

McNair, D.G. and Ditton, J., 'Does anybody know the cost of street lighting?', *The Lighting Journal*, Vol. 60, No. 5, October/November, pp. 299–300 (1995).

Nair, G. and Ditton, J., ' "In the dark, a taper is better than nothing": A one year follow-up of a successful streetlighting and crime prevention experiment'. *The Lighting Journal*, Vol. 59, No. 1, February, pp. 25–27 (1994).

Nair, G., Ditton, J. and Phillips, S., 'Environmental improvements and the fear of crime: The sad case of the "pond" area in Glasgow'. *British Journal of Criminology*, Vol. 33, No. 4, pp. 555–561 (1993).

Nair, G., McNair, D.G. and Ditton, J., 'Street lighting: Unexpected benefits to young pedestrians from improvement'. *Lighting Research and Technology*, 29, 143–149 (1997).

Nair, G., Ditton, J. and McNair, D.G., 'Crime in the dark: Studies in the relationship between streetlighting and crime in Scotland'. *Scottish Journal of Criminal Justice Studies*, Vol. 5, No. 1, pp. 75–107 (1999).

Painter, K.A., *The West Park Estate crime survey: An Evaluation of public lighting as a crime prevention strategy*. Cambridge: Institute of Criminology (1991).

Painter, K.A., 'The impact of street lighting on crime, fear and pedestrian street use'. *Security Journal*, 5, 116–124 (1994).

Painter, K.A., 'Street lighting, crime and fear of crime: A summary of research'. In T.H. Bennett (ed.) *Preventing Crime and Disorder: Targeting Strategies and Responsibilities* (pp. 313–351). Cambridge: Institute of Criminology, University of Cambridge (1996a).

Painter, K.A., 'The influence of street lighting improvements on crime, fear and pedestrian street use, after dark'. *Landscape and Urban Planning*, 35, 193–201 (1996b).

Painter, K.A. and Farrington, D.P., 'The crime reducing effect of improved street lighting: The Dudley project'. In R.V. Clarke (ed.) *Situational Crime Prevention: Successful Case Studies*, 2nd edn (pp. 209–226). Guilderland, NY: Harrow and Heston (1997).

Painter, K.A. and Farrington, D.P., 'Improved street lighting: Crime reducing effects and cost-benefits analyses'. *Security Journal*, 12, 17–32 (1999a).

Painter, K.A. and Farrington, D.P., 'Street lighting and crime: Diffusion of benefits in the Stoke-on-Trent project'. In K.A. Painter and N. Tilley (eds) *Surveillance of Public Space: CCTV, Street Lighting and Crime Prevention* (pp. 77–122) Monsey, NY: Criminal Justice Press (1999b).

Painter, K.A. and Farrington, D.P., 'The financial benefits of improved street lighting, based on crime reduction'. *Lighting Research and Technology*, 33, 3–12 (2001).

Pease, K., 'Crime prevention'. In M. Maguire, R. Morgan and R. Reiner (eds) *Oxford Handbook of Criminology*. Oxford: Clarendon (1997).

Pease, K., 'Repeat victimisation: Taking stock'. Crime Detection and Prevention Paper 92. London: Home Office Police Research Group (1998).

Pease, K., *Lighting and Crime*. Huddersfield University (1998).

Pease, K., 'A review of street lighting evaluations: Crime reduction effects'. In K.A. Painter and N. Tilley (eds) *Surveillance of Public Space: CCTV, Street Lighting and Crime Prevention* (pp. 47–76). Monsey, NY: Criminal Justice Press (1999).

Poyner, B. and Webb, B., 'Reducing theft from shopping bags in city centre markets'. In R.V. Clarke (ed.) *Situational Crime Prevention: Successful Case Studies*. New York: Harrow and Heston (1992).

Poyner, B. and Webb, B., 'What works in crime prevention: An overview of evaluations'. In R.V. Clarke (ed.) *Crime Prevention Studies 1*. Monsey, NY: Criminal Justice Press (1993).

Ramsey, M., 'Crime prevention: Lighting the way ahead'. *Home Office Research Bulletin*, 27, 18–20 (1989).

Ramsey, M. and Newton, R., 'The effect of better street lighting on crime and fear: A review'. Crime Prevention Unit Paper 29. London: Home Office (1991).

Shaftoe, H., 'Easton/Ashley, Bristol: Lighting improvements'. In S. Osborn (ed.) *Housing Safe Communities: An Evaluation of Recent Initiatives* (pp. 72–77). London: Safe Neighbourhoods Unit (1994).

Spelman, W., 'Criminal careers of public places'. In J.E. Eck and D. Weisburd (eds) *Crime and Place*. Monsey, NY: Criminal Justice Press and Police Executive Forum (1995a).

Spelman, W., 'Once bitten, then what? Cross-sectional and time course explanations of repeat victimisation'. *British Journal of Criminology*, 35, 366–383 (1995b).

Sternhill, R., 'The limits of lighting: The New Orleans experiment in crime reduction: Final impact evaluation report'. New Orleans, Louisiana: Mayor's Criminal Justice Coordinating Council (1997).

Tien, J.M., O'Donnell, V.F., Barnett, A. and Mirchandani, P.B., 'Street lighting projects: National evaluation program, Phase 1 Report'. Washington, DC: National Institute of Law Enforcement and Criminal Justice, U.S. Department of Justice (1979).

Tilley, N., 'Understanding car parks, crime and CCTV: Evaluation lessons from safer cities'. Crime Prevention Unit Paper 42. London: Home Office (1993).

Tyrpak, S., 'Newark high-impact anti-crime programme: Street lighting project interim evaluation report'. Newark, NJ: Office of Criminal Justice Planning (1975).

Vrij, A. and Winkel, F.W., 'Characteristics of the built environment and fear of crime: A research note on interventions in unsafe locations'. *Deviant Behavior*, 12, 203–215 (1991).

Weisburd, D., Green, L. and Ross, D., 'Crime in street level drug markets: A spatial analysis'. *Criminology*, 27, 49–67 (1994).

Welsh, B. and Farrington, D.P., 'Value for money? A review of the costs and benefits of situational crime prevention'. *British Journal of Criminology*, 38 (1998).

Wheeler, S., 'The challenge of crime in a free Society'. Presidential Commission on Law Enforcement and Administration of Justice. Washington, DC: US Government Printing Office (1967).

Wilson, J.Q. and Kelling, G.L., 'Broken Windows'. *Atlantic Monthly*, March, 29–38 (1982).

Wright, R., Heilweil, M., Pelletier, P. and Dickinson, K., *The Impact of Street Lighting on Crime*. Ann Arbor, Michigan: University of Michigan (1974).

General

Department of the Environment and Welsh Office, 'Making waste work: A strategy for sustainable waste management in England and Wales', HMSO, London, 1995.

Department of Environment, Transport and the Regions (DETR), 'A better quality of life: A strategy for sustainable development for the United Kingdom', HMSO, London, 1999.

Gardner, C., 'Re-valuing the pedestrian: A new approach to urban lighting', CIE Congress, Istanbul, September 2001.

Gardner, C., 'Strategic urban lighting plans in the UK: A preliminary balance sheet', CIE Congress, Vol. 2, p. 481, September 2001.

House of Commons Environmental Committee, 'Sustainable waste management inquiry', HMSO, London, 1998.

Lighting Design Partnership/Morris and Steadman Architects, 'Edinburgh lighting vision', Scottish Development Agency, 1990.

Twomey, S.A., Bohren, F.B. and Mergenthaler, J.L., 'Reflectance and albedo differences between wet and dry surfaces', *Applied Optics*, Vol. 25, No. 3, 1986.

World Commission on Environment and Development, 'Our Common Future (The Brundtland Report) – Report of the World Commission on Environment and Development', Oxford University Press, 1987.

Roads

Arend, H., Schwencke, K.R. and Zmech, D., 'Changes in traffic safety on urban motorways and arterial streets under wet conditions', *Strassen Verkehrs Technik*, Vol. 24, No. 4, 1980.

Boer, J.B. de and Knusden, B., 'The pattern of road luminance in public lighting', P63.17, p. 569, 1963.

Cobb, J., 'Road surface reflection characteristics'. Public Lighting Engineer's Conference, 1979, Paper No. 4.

De Clercq, G., 'Fifteen years of road lighting in Belgium', *International Lighting Review*, 1, 1985.

Dunbar, C., 'Necessary values of brightness contracts', IES, 1938.

Economopolous, I.A., 'Photometric parameters and visual performance in road lighting', University of Technology, Eindhoven, thesis, 1978.

Fisher, A.J. and Hall, R.R., 'Road luminances based on detection of changes of visual angle', *Lighting Research and Technology*, 8, 187–194, 1976.

Frederiksen, E. and Gudum, J., 'The quality of street lighting installations under changing weather conditions', *Lighting Research and Technology*, Vol. 4, No. 2, 1972.

Frederiksen, E. and Sorensen, K., 'Reflection classification of dry and wet road surfaces', *Lighting Research and Technology*, Vol. 8, No. 4, 1976.

Gallagher, V.P. and Meguire, P.G., 'Contrast requirements of urban driving', Transportation Research Board Special Report 156, Washington, 1975.

Haslegrave, C.M., 'Measurement of the eye heights of British car drivers above the road surface', Transport and Road Research Laboratory Report 494, 1979.

Lewin, Dr Ian, 'Lamp spectral effects at roadway lighting levels', *Lighting Journal*, Vol. 64/2, March/April, 1999.

McNair, D.G., 'The minimisation of the decrease in visual performance of the vehicle driver in wet conditions by adjustment of road lighting luminaires', thesis, 1995.

Mintsis, G. and Pitsiava-Latinopoulou, M., 'Traffic accidents with fixed roadside obstacles: a study of the Greek rural road network', Traffic Engineering and Control (GB), Vol. 31, No. 5, pp. 306–311, 1990.

Narisada, K., 'Influence of non-uniformity in road surface luminance of public lighting installations upon perception of objects on the road surface by car drivers', CIE XVIII Session, Barcelona, p71.17 (1971).

Sabey, B.E., 'Road surface reflection characteristics', Transport and Road Research Laboratory Report 490, 1972.

Sabey, B.E., 'Road accidents in darkness', Transport and Road Research Laboratory Report No. LR 536, 1973.

Scholz, I., 'Street lighting and motoring accidents', *International Lighting Review*, Vol. 29, p. 123, 1978.

Scott, P.P., 'The relationship between road lighting quality and accident frequency', Transport and Road Research Laboratory Report 929, 1980.

Sorenson, K., 'Description and classification of light reflection properties of road surfaces', Danish Illuminating Engineering Report 7, 1974.

Van Bommel, W.J.M., 'Principles of road lighting – requirements, recommendations and design'. Philips Engineering Report No. 39, 1978.

Van Bommell, W.J.M., 'Interrelation of road lighting quality criteria', *Lichtforschung*, Vol. 1, p. 10, 1979.

Waldram, J.M., 'The Revealing Power of Street Lighting Installations', *Transcripts of the Illumination Engineering Society* (London), Vol. 3, p. 173, 1938.

Walthert, R., 'The Influence of lantern arrangement and road surface luminance on subjective appraisal and visual performance in street lighting', CIE session, London, 1975, p75.60.

Vision

Adrian, W. and Eberbach, K., 'About the relation between contrast threshold and dimensions of background', *Optik*, Vol. 28, p. 132, 1968/69.

Davison, P.A. and Irving, A., 'Survey of visual acuity of drivers', Transport and Road Research Laboratory Report 945, 1980.

Fisher, J.M. and Christie, A.W., 'A note on disability glare', *Vision Research*, Vol. 5, p. 565, 1965.

Lewin, Dr Ian, 'Visibility factors in outdoor lighting design (Parts 1 and 2)', *Lighting Journal*, Vol. 64/6, November/December 1999 and Vol. 65/1, January/February 2000.

Loe, D., Mansfield, K. and Rowlands, T., 'Appearance of the lit environment and its relevance in lighting design: Experimental study', *Lighting Research and Technology*, Vol. 26/3, 1994.

Rea, M.S. and Ouellette, M.J., 'Visual performance using reaction times', *Lighting Research and Technology*, Vol. 20, No. 4, pp. 139–153, 1988.

Guides and standards

ILE technical reports

**	Guidance Notes for the Reduction of Light Pollution (2000).
**	Domestic Security Lighting, Friend or Foe (2001).
**	Lighting the Environment – A guide to good urban lighting (1999).
**	The use of remote monitoring and switching technology in street lighting services (1999).
CP1	Code of practice for electrical safety in highway electrical operations (2003).
CP2	Lasers, Festival and Entertainment Lighting Code (1995).
TR5	Brightness of Illuminated Advertisements (2001).
TR7	High mast lighting (second edition) (1996) (Superceded).
TR7	High masts for lighting and CCTV (2000 edition) (2000).
TR9	Protective coatings for steel lighting columns (1980).
TR12	Lighting for pedestrian crossings (1997).
TR18	The planned replacement of lighting columns (1988).
TR19	The cost effectiveness of luminaire cleaning (1989).
TR22	Lighting columns and sign posts: Planned inspection regime (second edition) (2002).
TR23	Lighting for cycle tracks (1998).
TR24	A Practical Guide to the Development of a Public Lighting Policy for Local Authorities (1999).
TR25	Lighting for Traffic Calming Features (2002).
TR26	Painting of Lighting Columns (2003).

SLL Lighting Guides

LC1	Code for lighting (2002).
LG1	The industrial environment (1989).

LG4   Sports (1990, addendum 2000).
LG6   The outdoor environment (1992).
          Lighting for Railway Premises – Good Practice Guide (1992)

ILE and SLL joint publications

Guide to fibre optic and remote source lighting (2002).
Lighting the Environment – A guide to good urban lighting (1995).

CIE publications

1xx   Guide on the limitation of the effects of obtrusive light from outdoor
          lighting installations (Draft 2001).
01   Guide lines for minimizing Urban Sky Glow near Astronomical
          Observatories (1980).
13.3   Method for measuring and specifying colour rendering of light
          sources (Includes disk D008) (1995).
17.4   International lighting vocabulary, 4th edition (1987) (Joint
          publication IEC/CIE).
18.2   The basis of physical photometry, 2nd edition (1983).
19/2.1   An analytic model for describing the influence of lighting parameters
          upon visual performance, Volume 1, Technical foundations (1981).
19/2.2   An analytic model for describing the influence of lighting parameters
          upon visual performance, Volume 2, Summary and application
          guidelines (1981).
23   International recommendations for motorway lighting (1973).
30   Calculation and Measurement of Luminance and Illuminance in
          Road Lighting (1976).
31   Glare and uniformity in road lighting installations (1976).
32   Lighting in situations requiring special treatments (in road lighting)
          (1977).
33   Depreciation of installations and their maintenance (in road lighting)
          (1977).
34   Road lighting lantern and installation data: Photometrics,
          classification and performance (1977).
38   Radiometric and photometric characteristics of materials and their
          measurement (1977).
41   Lighting as a true visual quantity: Principles of measurement (1978).
42   Lighting for tennis (1978).
43   Photometry of floodlights (1979).
44   Absolute methods of reflection measurements (1979).
45   Lighting for ice sports (1979).
46   A review of publications on properties and reflection values of
          material reflection standards (1979).
47   Road lighting for wet conditions (1979).
53   Methods of characterizing the performance of radiometers and
          photometers (1982).
57   Lighting for football (1983).

61   Tunnel entrance lighting: A survey of the fundamentals for determining the luminance in the threshold zone (1984).

62   Lighting for swimming pools (1984).

66   'Road Surfaces and Lighting', C.I.E./P.I.A.R.C. joint technical report (1984).

67   Guide for photometric specifications and measurements of sports lighting installations (1986).

69   Methods of characterizing illuminance meters and luminance meters: Performance, characteristics and specification (1987).

70   The measurement of absolute luminous intensity distributions (1987).

73   Visual aspects of road markings (Joint technical report CIE/PIARC) (1988).

81   Mesopic photometry: History, special problems and partial solutions (1989).

83   Guide for the lighting of sports events for colour television and film systems (1989).

84   Measurement of luminous flux (1989).

86   CIE 1988 2° spectral luminous efficiency function for photopic vision (1990).

88   Guide for the lighting of road tunnels and underpasses (1990).

92   Guide for floodlighting (1992).

93   Road lighting as an accident countermeasure (1992).

94   Guide for floodlighting (1993).

95   Contrast and visibility (1992).

100   Fundamentals of the visual task of night driving (1992).

101   Parametric effects in colour-difference evaluation (1993).

109   A method of predicting corresponding colours under different chromatic and illuminance adaptations (1994).

112   Glare evaluation system for use within outdoor sports and area lighting (1994).

113   Maintaining night-time visibility of retroreflective road signs (1995).

115   Recommendations for the lighting of roads for motor and pedestrian traffic (1995).

121   The photometry and goniophotometry of luminaires (1996).

126   Guidelines for minimizing Skyglow (1997).

127   Measurements of LEDs (1997).

128   Guide to the lighting of open-cast mines (1998).

129   Guide for lighting exterior work areas (1998).

130   Practical methods for the measurement of reflectance and transmittance (1998).

131   The CIE 1997 interim colour appearance model (simple version) CIECAM97s (1998).

132   Design methods for lighting roads (1999).

136   Guide to the lighting of urban areas (2000).

139   The influence of daylight and artificial light on diurnal and seasonal variations in humans – a bibliography (2001).

140   Road lighting calculations (2000).

142   Improvement to industrial colour difference evaluation (2001).

143   International recommendations for colour vision requirements for transport (2001).

144   Road surface and road marking reflection characteristics (2001).
145   The correlation of models for vision and visual performance (2002).
146   CIE equations for disability glare (2002).
147   Glare from small, large and complex sources (2002).
150   Guide on the limitation of the effects of obtrusive light from outdoor lighting installations (2003).
151   Spectral weighting of solar ultraviolet radiation (2003).
153   Report on an intercomparison of measurement of the luminous flux of high pressure sodium lamps (2003).
154   The maintenance of outdoor lighting systems (2003).

Bulletin No. 30, 'Statement from the Commission Internationale de l'Éclairage on Vehicle front lighting used on urban traffic routes' (1976).

CIE publications on diskettes

D001   Disc version of CIE photometric and colorimetric data (tables from Publ. 18.2, 86, ISO 10526/CIE S005 and ISO/CIE 10527) (1988).
D002   Disc version of CIE colorimetric and colour rendering data (Publ. 13.3 and 15.2 tables) (1995).
D007   A computer program implementing the 'Method of predicting corresponding colours under different chromatic and illuminance adaptations' described in Publication CIE 109-1994.

CiE Standards

ISO 10526/CIE S005: Joint ISO/CIE Standard: CIE Standard Illuminants for Colorimetry (1999).
ISO/CIE 10527: Joint ISO/CIE Standard: Colorimetric Observers, 1991 (S002, 1986).

CIE: Draft Standards

CIE Draft Standard DS 015:2002: Lighting of Work Places – Outdoor Work Places.

CIE: Proceedings of conferences and symposia

x008   Urban Sky Glow, a Worry for Astronomy (1994).
x019   Proceedings of three CIE Workshops on Criteria for Road Lighting (2001).
x023   Proceedings of 2 CIE Workshops on Photometric Measurement Systems for Road Lighting Installations (Liege 1994, Poitiers 1996) (2002).

Lighting Industry Federation Technical Statements

No. 8   Ultraviolet Radiation and Health (1995).
No. 10   The Handling and Disposal of Lamps (1998).

No. 11   Compact Fluorescent Lamps, Power Factor and Harmonic Content
                (1995).
No. 15   European Voltage Harmonisation (1995).
No. 19   Seasonal Affective Disorder (1995).
No. 22   ENEC Mark (1997).

Countryside Commission/DOE

Lighting in the Countryside: Towards Good Practice (1997) (Out of Print).

Department of Transport

Road Lighting and the Environment (1993) (Out of Print).

Sports publications

DoE Good Practice Guide 223.
Sport England Guidance Notes for Floodlighting.
Lawn Tennis Association.
National Playing Fields Association Sports Lighting Guide.
There are specific requirements by FIFA and UEFA for International matches and
competitions.
Many domestic leagues and competitions have specific requirements e.g. FA Cup,
FA trophy and Football League.

British Standards

BS 873   Part 3: Specification for Internally Illuminated Bollards.
BS 5489–1   Code of Practice for the Design of Road Lighting – Lighting for
                Roads and Public Amenity Areas (2003).
BS 5489–2   Code of Practice for the Design of Road Lighting – Lighting for
                Tunnels (2003).
BS 5493   Code of Practice for Protective Coating of Iron and Steel Structures
                against Corrosion (1977) (withdrawn).
BS 5649   Lighting Columns (withdrawn).
BS 5649   Part 2/EN 40-2, Lighting Columns. Dimensions and Tolerances
                (1978) (withdrawn).
BS 5649   Part 3, Lighting Columns. Specification for Materials and Welding
                Requirements (1982) (withdrawn).
BS 5649   Part 4/EN 40-4, Lighting Columns. Recommendations for Surface
                Protection of Metal Lighting Columns (1982) (withdrawn).
BS 5649   Part 5/EN 40-5, Lighting Columns. Specification for Base
                Compartments and Cableways (1982) (withdrawn).
BS 5649   Part 6/EN 40-6, Lighting Columns. Specification for Design Loads
                (1982) (withdrawn).
BS 5649   Part 7/Lighting Columns. Method for Verification of Structural
                Design by Calculation (withdrawn).

BS 5649   Part 8/EN 40-8, Lighting Columns. Method for Verification of
                   Structural Design by Testing (1982) (withdrawn).
BS 5649   Part 9/EN 40-9, Lighting Columns. Specification of Special
                   Requirements for Reinforced and Prestressed Concrete Lighting
                   Columns (1982) (withdrawn).
BS 6399   Loading for Buildings, Part 2, Code of Practice for Wind Loads
                   (1997).
BS 7430   Code of Practice for Earthing (1998).
BS 7671   Requirements for Electrical Installations (2001).

European Standards

BS EN 40-1   Lighting Columns – Definitions and Terms (1992).
BS EN 40-2   Lighting Columns – General requirements and dimensions (2004).
BS EN 40-3-1   Lighting Columns – Design and Verification – Specification for
                   Characteristic Loads (2000).
BS EN 40-3-2   Lighting Columns – Design and Verification – Verification by
                   Testing (2000).
BS EN 40-3-3   Lighting Columns – Design and Verification – Verification by
                   Calculation (2003).
BS EN 40-4   Lighting Columns – Concrete Columns (pending).
BS EN 40-5   Lighting Columns – Requirements for Steel Lighting Columns
                   (2002).
BS EN 40-6   Lighting Columns – Requirements for Aluminium Lighting
                   Columns (2002).
BS EN 40-7   Lighting Columns – Specification for Composite Lighting Columns
                   (2002).
BS EN 755   Aluminium and Aluminium Alloys. Extruded Rod/Bar, Tube and
                   Profiles (1997).
BS EN 1706   Aluminium and Aluminium Alloys. Castings. Chemical
                   Composition and Mechanical Properties (1998).
BS EN 10025   Specification for Hot Rolled Products of Non-Alloy Structural
                   Steels and their Technical Delivery Conditions (1993).
BS EN 10088   Stainless Steels (1995).
BS EN 10210   Hot Finished Structural Hollow Sections of Non-Alloy and Fine
                   Grain Structural Steels (1994).
BS EN 10219   Cold Formed Welded Structural Hollow Sections of Non-Alloy
                   and Fine Grain Steels (1997).
BS EN 12193   Light and Lighting – Sports Lighting (1999).
BS EN 12665   Lighting applications: Basic Terms and Criteria for Specifying
                   Lighting Requirements (2002).
BS EN 12767   Passive Safety of Support Structures for Road Equipment –
                   Requirements and Test Methods (2000).
BS EN 13201-2   Road lighting – Performance requirements (2003).
BS EN 13201-3   Road Lighting – Calculation of Performance (2003).
BS EN 13201-4   Road lighting – Methods of Measuring Light Performance (2003).
BS EN 50102   Degrees of Protection Provided by Enclosures for Electrical
                   Equipment against External Mechanical Impacts (IK code).

BS EN 60309   Plugs, Socket-outlets and Couplers for Industrial Purposes, Part 1: 1988 – General requirements.

BS EN 60309   Plugs, Socket-outlets and Couplers for Industrial Purposes, Part 2: 1988 – Dimensional Interchangeability Requirements for Pin and Contact Tube Accessories.

BS EN 60529   Specification for Degrees of Protection Provided by Enclosures (IP Code) (1992).

BS EN 60598   Part 1 Covers Luminaires Generally and is Entitled 'General Requirements and Tests' (2004).

BS EN 60598   Part 2 is Issued as a Number of Sections each Dealing with a Particular Type of Luminaire; Part 2 refers to sections of Part 1 Specifying the Extent to which it is Applied, and any Additional Requirements.

BS EN 60598   Part 2.3 Covers Luminaires for Road and Street Lighting (2003).

BS EN 60598   Part 2.5 Floodlights (1998).

EN/IEC 61347-1   Lamp Control Gear: General and Safety Requirements (2003).

EN/IEC 61347-2-1   Ignitors (2003).

EN/IEC 61347-2-2   AC Supplied Electronic Step Down Convertor (2003).

EN/IEC 61347-2-3   AC Supplied Electronic Ballast for Fluorescent Lamps (2003).

EN/IEC 61347-2-4   DC Supplied Electronic Ballasts for General Lighting (2003).

EN/IEC 61347-2-5   DC Supplied Electronic Ballast for Public Transport Lighting (2003).

EN/IEC 61347-2-6   DC Supplied Electronic Ballast for Aircraft Lighting (2003).

EN/IEC 61347-2-7   DC Supplied Electronic Ballast for Emergency Lighting (2003).

EN/IEC 61347-2-8   Ballast for Fluorescent Lamps (2003).

EN/IEC 61347-2-9   Ballasts for Discharge Lamps (2003).

EN/IEC 61347-2-10 Electronic Invertors and Convertors for High Frequency Operation of Cold Start Tubular Discharges Lamps (neon tubes) (2003).

International Standards Organisation

BS EN ISO 1461   Hot Dip Galvanised Coatings on Fabricated Iron and Steel Articles – Specifications and Test Methods (1999).

BS EN ISO 7091   Washers, Normal Series (2000).

BS EN ISO 7093   Washers, Large Series (2000).

BS EN ISO 9000   Quality Management Systems. Fundamentals and Vocabulary (2000).

BS EN ISO 12944   Paints and varnishes. Corrosion Protection of Steel Structures by Protective Paint Systems (1998).

BS MA 29   Specification for Steel Wire Rope and Strand for Yachts (1982).

DD ENV 1991-2-4   Action on Structures. Wind Actions (together with United Kingdom national application document) (1997).
PD IEC TS 61231   International Lamp Coding System (ILCOS) (1999).

British Standard Codes of Practice

CP3, Chapter V, Part 2, Wind Loads, 1972. British Standards (obsolescent, but still used).

BSI Technical reports

PD6484 (1979), confirmed March 1990, Commentary on Corrosion at Bimetallic Contact and its Alleviation.

Department of Transport

Specification for Highway Works: Volume 1: 2002, Series 1300.
Notes for Guidance on Specification for Highway Works: Volume 2: 2002, Series 1300.
Design Manual for Roads and Bridges: Volume 2: Section 2: Part 1 BD26/99 – Design of Lighting Columns.

Other

Environment Agency Guidance Note SWEN 047 Special Waste Explanatory Notes for Lamps Containing Mercury.
Environment Agency Guidance Note SWEN 047A Special Waste Explanatory Notes for lamps Containing Sodium.
Environment Agency of Japan, 'Guideline for Light Pollution: Aiming for Good Lighting (1998).
Italian Standard UNI 10819, 'Outdoor Lighting Installations: Requirements for Limiting the Upward Flux Scattering' (1999).

Web sites

Lighting

www.darksky.org	The International Dark-Sky Association site contains a good description of light pollution, how to avoid it, legislation enacted to avoid it, and a host of other useful snippets.
www.cibse.org	Chartered Institution of Building Services Engineers.
www.cie.co.at	The Commission Internationale de l'Eclairage.

(Continued)

Lighting

www.homeoffice.gov.uk/rds/ horspubs1.html	This site allows access to a download of the reports Home Office Research Study 251: Effects of improved street lighting on crime: a systematic review; and, Home Office Research Study 252: Crime prevention effects of closed circuit television: a systematic review.
www.ile.org	The premier web site for lighting practitioners in the UK. For members, a chat room ensures access to someone prepared to discuss virtually anything.
www.iesna.org	Illuminating Engineering Society of North America (IESNA).
www.lif.co.uk	Lighting Industry Federation.
www.lightingacademy.org	An Italian site with comprehensive explanations of lighting principles and practice.
www.lta.org.uk	This site includes the Lawn Tennis Association recommendations for lighting of tennis courts.
www.mastercolour-city.com	A Philips Lighting site devoted to ceramic metal halide lamps.
www.outdoorlamps.philips.com	A Philips Lighting site about outdoor lamps.
www.petrolstationlighting.com	A Philips Lighting site about petrol station lighting.
www.planning.odpm.gov.uk/ litc/index.htm	This address allows access to a download of the out of print document 'Lighting in the countryside'.
www.philipslampsandgear.co.uk	An excellent site with a number of LIF technical statements on issues from lamp disposal to seasonal affective disorder.
www.ql-lighting.com	A Philips Lighting site devoted to QL lamps.
www.sustainalite.co.uk	SustainaLite Scheme.
www.tut.fi/cie4	The home page of CIE Division 4, lighting and signalling for transport.

Others

www.bsi-global.com	British Standards Institute.

www.cctvconsultant.co.uk	A site that contains lots of useful information about CCTV systems.
www.cenelec.org	European Committee for Electrotechnical Standardisation.
www.ciwm.org	Chartered Institution of Waste Management.
www.cpre.org.uk	Campaign to protect Rural England.
www.dft.gov.uk	Department for Transport.
www.dti.gov.uk	Department of Trade and Industry.
www.drdni.gov.uk	Department for Regional Development (Northern Ireland).
www.engc.org.uk	The Engineering Council.
www.environment-agency.gov.uk	The Environment Agency.
www.etechb.co.uk	The Engineering and Technology Board.
www.highways.gov.uk	Highways Agency.
www.homeoffice.gov.uk	Home office.
www.hse.gov.uk/hsehome.htm	Health and Safety Executive.
www.iec.ch	International Electrotechnical Commission.
www.iee.org	Institution of Electrical Engineers.
www.lightswitch.co.uk	Energy saving trusts.
www.nacro.org.uk	A charity dedicated to the reduction of crime.
www.naturalstep.org.uk	Hints for successful organisation planning for sustainability.
www.nds.coi.gov.uk	Central Office of Information.
www.ni-assembly.gov.uk	Northern Ireland Assembly.
www.odpm.gov.uk	Office of the Deputy Prime Minister (UK).
www.ofgem.gov.uk	Office of the gas and electricity markets.
www.pacts.org.uk	Parliamentary Advisory Council for Transport Safety.
www.parliament.uk/commons/hsecom.htm	House of Commons (UK).
www.patent.gov.uk	The United Kingdom patent office.
www.rethinkingconstruction.org.uk	Rethinking Construction: A United Kingdom government and industry partnership.
www.scottish.parliament.uk	Scottish Parliament.
www.sepa.org.uk	Scottish Environmental Protection Agency.
www.wales.gov.uk	Welsh Assembly.

Legislation

The Health and Safety at Work etc. Act.
The Management of Health and Safety at Work Regulations.
The Provision and Use of Work Equipment Regulations.
Lifting Equipment and Lifting Operations Regulations.
The Construction, Design and Management Regulations.
The Construction (Health, Safety and Welfare) Regulations.
The Control of Substances Hazardous to Health (COSHH) Regulations.
The Electricity at Work Regulations.
The Workplace (Health, Safety and Welfare) Regulations.
Special Waste Regulations.
The Building Regulations Part L.
The Low Voltage Directive (LVC), 72/23/EEC.

Code of Practice for In Service Inspection and Testing of Electrical Equipment.
Electrical Maintenance.
On Site Guide.
Guidance Note 1 – Selection and Erection.
Guidance Note 2 – Isolation and Switching.
Guidance Note 3 – Inspection and Testing.
Guidance Note 4 – Protection against Fire.
Guidance Note 5 – Protection against Electric Shock.
Guidance Note 6 – Protection against Over-current.
Guidance Note 7 – Special Locations.

UK Bodies with responsibility for waste management

Scottish Environmental Protection Agency (SEPA)
In Scotland, monitoring and enforcement is carried out by SEPA. Its main aim is to provide an efficient and integrated environmental protection system for Scotland that will both improve the environment and contribute to the Scottish Ministers' goal of sustainable development.

Northern Ireland Environment and Heritage Service (NIEHS)
NIEHS is the body charged with environmental management in Northern Ireland.

Environment Agency (EA)
In England and Wales the EA carries out monitoring and enforcement of environmental issues as required.

The Chartered Institution of Waste Management (CIWM)
The CIWM is a professional body which represents over 5000 waste management professionals predominantly in the UK, but also overseas. This institution sets the professional standards for individuals working in the waste management industry and has various grades of membership determined by education, qualification and experience. It also organises seminars, exhibitions, training events and technical publications to keep its members abreast of current issues in waste management.

The objectives of the CIWM are as follows:

- Advancing the scientific, technical and practical aspects of wastes management for the safeguarding of the environment.
- Promoting education, training, research and the dissemination of knowledge in all matters of waste management.
- Striving to achieve and maintain the highest standards of practice, competence and conduct by all its members.

Index

Note: Bold locators refer to detailed description of the entry.